D1300450

Books of Merit

The Quantum Ten

The QUANTUM TEN

A STORY *of* PASSION TRAGEDY, AMBITION AND SCIENCE

SHEILLA JONES

THOMAS ALLEN PUBLISHERS

TORONTO

Library and Archives Canada Cataloguing in Publication

Jones, Sheilla, 1954–
 The quantum ten : a story of passion, tragedy, ambition and science / Sheilla Jones.

Includes bibliographical references.
ISBN 978-0-88762-331-8

1. Physicists—Biography.
2. Physics—History.
3. Quantum theory—History.
4. Solvay Conference on Physics (5th : 1927 : Brussels).
I. Title.

QC15.J65 2008   530.092'2   C2008-900270-9

Editor: Janice Zawerbny
Jacket and text design: Gordon Robertson
Jacket image: Emilio Segre Archives
Author photograph: Fred Elcheshen

Published by Thomas Allen Publishers,
a division of Thomas Allen & Son Limited,
145 Front Street East, Suite 209,
Toronto, Ontario M5A 1E3 Canada

www.thomas-allen.com

 ONTARIO ARTS COUNCIL
CONSEIL DES ARTS DE L'ONTARIO

 Canada Council
for the Arts

The publisher gratefully acknowledges the support of
the Ontario Arts Council for its publishing program.

We acknowledge the support of the Canada Council for the Arts, which last
year invested $20.1 million in writing and publishing throughout Canada.

We acknowledge the Government of Ontario through the Ontario
Media Development Corporation's Ontario Book Initiative.

We acknowledge the financial support of the Government of Canada through the
Book Publishing Industry Development Program (BPIDP) for our publishing activities.

12 11 10 09 08   1 2 3 4 5

Printed and bound in Canada

To Dr. James Burns

This narrative is filled with fascinating characters whose personal stories have been brought to life by their biographers, in particular: David C. Cassidy, Paul Carter, Charles P. Enz, Philipp Frank, Roger Highfield, Banesh Hoffman, Martin J. Klein, Helge Kragh, Thomas Levenson, David Lindley, Walter Moore, Abraham Pais and Nancy Thorndike Greenspan. My interpretation of the tumultuous development of quantum physics is greatly enriched by their earlier efforts.

In particular, I'd like to thank Martin J. Klein for his invaluable insights and critique of much of the manuscript during the writing of this book, as well as for sharing some of his unpublished research on the distressing final hours before Paul Ehrenfest's suicide. Paul Halpern and Frans van Lunteren were also most helpful in tracking down additional details on Ehrenfest's last years.

This book is very much a product of the enthusiasm of my agent Rick Broadhead, editor Janice Zawerbny, and my biggest fan and resident grammarian, Jim Burns. Thank you, all.

CONTENTS

The Quantum Ten

ALBERT EINSTEIN (1879–1955)

A lone wolf and world celebrity, who stood his ground over
what should be kept from the old physics and what could be
abandoned to create a new physics.

NIELS BOHR (1885–1962)

An obsessive and pressured man, running his own physics
institute in Copenhagen, raising a large family and getting
left behind by the new mathematical physics.

PAUL EHRENFEST (1880–1933)

An intense physicist with a debilitating streak of self-doubt
who could rarely see the valuable gift he offered to physics,
and a passionate friend to both Einstein and Bohr.

MAX BORN (1882–1970)

An anxious hypochondriac, madly in love with his difficult wife,
and determined to build his physics institute into a world
leader in mathematical physics.

ERWIN SCHRÖDINGER (1887–1961)

An enthusiastic womanizer who was constantly getting distracted from his innovative work in science by the pleasures of
the flesh.

WOLFGANG PAULI (1900–1958)

> A moody young man with a fondness for nightclubs and wine who would sink into a depression when he failed to live up to the exalted standards he set for himself in his work.

LOUIS DE BROGLIE (1892–1987)

> A French aristocrat whose mother felt that his seemingly frustrated efforts in the new science meant her youngest son, by the standards of the de Broglie dynasty, was a failure.

WERNER HEISENBERG (1901–1976)

> A handsome, ambitious young man whose self-confidence was continually undermined by his father's anxious concern that he really wasn't smart enough to be a physicist.

PAUL DIRAC (1902–1984)

> A tall, slender Englishman, reputed to have a vocabulary of "yes," "no" and "I don't know," who let his inventive mathematics speak for him.

PASCUAL JORDAN (1902–1980)

> A passionate young German nationalist with a challenging speech impediment and a powerful grasp of mathematical physics.

Chapter One

THE REGRESSION OF SCIENCE

Science . . . is part and parcel of our knowledge and obscures
our insight only when it holds that the understanding given by
it is the only kind there is.

CARL JUNG

Theoretical physics is in trouble. At least that's the impression you'd get from reading a spate of recent books on the continued failure of the reigning star of theoretical physics—superstring theory—to resolve the eighty-year-old problem of unifying the classical and quantum worlds.

Over the past three decades, a whole superstring industry in academia has grown out of the attempts by theoretical physicists to capture the elusive Holy Grail of physics by, at long last, coming up with a way of integrating the two completely incompatible sets of physics rules for our universe. There are careers and reputations at stake, not just for the leading string theorists and their entourages of post-docs and graduate students, but also for physics departments that have channelled much of their funding and talent into strings.

It's not surprising, then, that a growing unease in the physics community about the validity of string theory as a viable research enterprise is turning into a rather testy exchange between those with a vested interest in keeping it going and those who think it's an idea that has run its course—and consumed quite enough of the limited funding available for theoretical research. The issue is being debated in science journals and in the mainstream media, as well as in online physics forums and blogs. At times, the level of discussion between defenders of string theory and its challengers has descended to a decidedly childish and even offensive level. When some scientists feel their careers and favoured ideas are threatened, the need to defend and protect their interests can far outweigh the romantic ideal of scientific research as a pursuit of truth, unsullied by human emotion and personal agendas.

This kind of angry and defensive push-pull in scientific inquiry is nothing new. The same thing happened eight decades ago during the creation of quantum physics. That's when a set of physics rules was established for the quantum or microscopic world, but it was a set of rules that was completely incompatible with the existing set for the classical or macroscopic world. Albert Einstein objected to the rules for quantum physics while they were being developed precisely because they appeared to preclude any means of reconciliation with the classical rules upon which his generalized theory of relativity was based. String theorists are still trying to finish what Einstein started eighty years ago, and with about as much success as Einstein had.

If, after such a long time, all the smart men and all the smart women who work in physics have not been able to reconcile the two sets of rules for the universe, it's natural to wonder if one—or both—of the sets might just be wrong. But suggesting that either of the two most successful, experimentally verified models of physics might be wrong is tantamount to goring a sacred cow. Just as questioning the validity of string theory can raise the hackles of its supporters, so can questioning relativity or quantum physics.

A discomfort with the strange and seemingly inexplicable properties of the quantum has been bubbling under the surface of quantum physics since its creation. And at times the discomfort has boiled over into heated debate between physicists over whether there's something fundamentally wrong with quantum physics or whether it's just fine the way it is. Indeed, there have been more than a few public yelling matches between leading physicists, shouting at each other across conference rooms, about whether it's time to revisit the development of quantum physics and see if, just maybe, there is another way of looking at the quantum world.

Perhaps it *is* time, particularly in light of an emerging ideological trend in string theory. The key weakness of string theory is that it is not experimentally verifiable or falsifiable—although some of its leading proponents will posit that it could be tested if, for instance, we could build a particle accelerator the size of the Milky Way Galaxy[1]—and, being untestable, string theory can be considered a highly elaborate mathematical exercise. Physicist and string theory popularizer Brian Greene has admitted that unless string theory can make predictions that are subject to experimental verification, "it will be no more relevant than an elaborate game of Dungeons and Dragons."[2] Since it's highly unlikely humans are ever going to build a galactic-scale particle accelerator, a few of the world's leading physicists have begun toying with the idea that the problem of string theory's non-testability might be eliminated by dropping experimental testing as a condition for determining the validity of the string proposal.

It may seem rather bizarre to propose that research in physics could be advanced by dropping the actual physical testing, but it's not a new idea. In the world of the ancient philosophers, rational and logical thinking was all that was required to understand the physical world; to conduct experiments was to sully the art of pure intellectual reasoning. That particular viewpoint served to stall scientific progress for about two thousand years. For the past five centuries, however, scientific progress has flourished with the help of experimental verification.

It used to be that modern physics, the most fundamental of sciences, progressed by a quite workable marriage of philosophical theory, mathematically based physics rules and experimentation. When Einstein began working on his theory of general relativity, he started with a thought experiment and a hypothesis. He imagined that gravity was actually the effect of mass curving space and curving the path of light as well. He could visualize it. Einstein then set about finding the mathematics with which to calculate the curvature. He wasn't very good at math, so he relied on a couple of friends who were also physicists to give him a hand. By 1917 they had worked out all the kinks in the math, but it wasn't until the solar eclipse of 1919 that Sir Arthur Eddington and his crew gathered the supporting evidence for Einstein's hypothesis. Eddington and others were able to show that light from stars was indeed caused to curve as it passed by the huge gravitational mass of the sun.

The process Einstein followed is one quite familiar to high school or fifth form science students. He used a long-standing method of conducting science: form a hypothesis that predicts nature will behave a certain way, find an equation suitable for making predictions, and then conduct experiments to validate or falsify the hypothesis. Once Einstein's prediction that mass curved space was borne out by experimental validation, his hypothesis became a theory.

But not long after Einstein's high-profile success with general relativity—it had made him a world celebrity—the scientific method began to change. The cause? The discovery of a strange new atomic realm with bizarre rules that were impossible to visualize.

The quantum world of the atom operates by rules that are physically impossible in our everyday world. It's a very strange notion that everything in the ordinary world is made up of "quantum stuff," such as electrons, light waves and atoms, and yet there is no way to make the macroscopic and microscopic worlds operate under a single set of rules. Despite the best efforts of some of the greatest scientific thinkers of the time, the quantum revolution of the mid-1920s failed to produce a philosophical world view capable of combining the two disparate worlds into one sensible theory. If anything, the scientists who developed quantum physics created the very unification problem that string theorists are still trying to resolve.

At this point, it's worth adding some clarity to the terms physicists use to talk about quantum science. They often treat the phrases "quantum mechanics," "quantum theory" and "quantum physics" as if they were interchangeable, which they are not. Throughout this book, "quantum mechanics" is defined as the set of rules for *how* the physics and mathematics are used to make testable predictions; these rules have been used to unparalleled fruitfulness since their development in the 1920s. "Quantum theory" is defined as the explanation for *why* the quantum world behaves the way it does; this exercise is still fraught with controversy. "Quantum physics" is the whole package—the mechanics and the theory.

By the 1930s, most physicists simply abandoned the need for a philosophical theory of quantum physics, mainly because no one had been able to come up with an explanation that made any kind of sense of the strange quantum world. The rush of discoveries using just the mathematics-based quantum mechanics and experimental testing produced new fundamental laws to explain atomic structure and identify the kinds of particles that make up the universe—along with Nobel prizes for scientists at the forefront of discovery. What need for a unified, philosophical world view when the creative use of mathematics produced such marvellous results?

By the 1940s, not only was philosophy no longer considered a necessary part of science, it was seen as antithetical to science. (To this day, the ultimate put-down in physics is to call someone's work "philosophy.") Truth could be established using only mathematics and experiments, with the logic of mathematics driving discovery. This fit well with the dominant philosophy of the time, particularly the one advocated by the Vienna Circle which flourished in the 1920s and 1930s.

The goal of the European philosophers, scientists and mathematicians who made up the Vienna Circle was to develop a unified science where true knowledge, as they defined it, could be obtained from the evidence of the senses alone (what could be observed, counted or otherwise measured) and by the use of analytical logic, particularly the kind used in mathematics. There was no place for gods or metaphysics or any other kind of airy-fairy theories here. Indeed, the point of positivist philosophy was to create a world view "positively" free of metaphysics, mysticism and the supernatural, and to anchor knowledge and truth in something verifiable (via experimental testing) or logically consistent (mathematical). This philosophy was called *logical positivism*. The popular philosophical model promulgated by the Vienna Circle fit the math-driven quantum physics like a glove and helped to make the strange new science more palatable. However, the remarkable experimental success of quantum mechanics masked the empty place where a coherent theory should have been.

By the 1960s, the fruitfulness of the math-driven theory and experimentation began to wane. There have been no new fundamental laws of nature discovered since the 1970s, and there is still no math-driven theory that can reconcile the quantum and classical worlds.

The string theory posits that the universe is not made up of elementary particles such as electrons, quarks and neutrinos, but comprises incredibly small "strings" or tiny rips in the space-time continuum that operate in ten or eleven dimensions. Just try to visualize that! Theoretical physicists have also tried to unify classical gravity with the quantum world through a field of study called quantum gravity, but it too has foundered on the fact that gravity is so weak at the atomic level that it is outside the range of experimental testing.

Most math-driven theorists in physics have long since parted company with the experimentalists, so much so that there is now a whole generation of theoretical physicists who know nothing about how to design experiments to test their hypotheses—and, moreover, feel no need to know. Dutch Nobel laureate Gerard 't Hooft has recently argued that scientific progress can continue without experiments if theorists carefully apply rigorous logic. For instance, there is no possible way to test theories that involve the incredibly tiny atomic measurement called a Planck length.* This means, according to 't Hooft, "the only instrument we can use to study the Planck length in detail, is our minds." But 't Hooft is also mindful of how important it has been to be able to use experimental verification to distinguish between competing theories. "Unfortunately," he says, "history has also shown how easily we can be misled into wrong theories, and how important it was, throughout the ages, that we could ask Nature to settle our disputes."[3]

American Nobelist Steven Weinberg has suggested that this could mark a "heroic age when theorists cut themselves temporarily free from their experimental underpinnings" and make use of "pure theoretical reasoning to develop a unified theory of all the phenomena of nature."[4]

But isn't that how philosophers such as Pythagoras, Plato and Archimedes did science some 2,500 years ago? They placed no value on experiments, but rather relied on carefully thinking through ideas until they arrived at a logically consistent explanation for how the natural world worked. Pythagoras founded his own mathematics-based religion, seeking the divine principles governing the universe through pure thought and analytical reasoning. Plato, for his part, condemned the use of mechanical measuring devices in geometry because it reduced the "incorporeal and intellectual" art to a "vulgar handicraft." Archimedes did do some experimentation but nonetheless held that it was quite inappropriate for philosophers to make practical use of science in the mundane world.[5]

By limiting themselves to intellectual reasoning, the scientists of ancient times limited themselves to a single means of determining truth and reality. Their scientific method relied on a one-trick pony. It wasn't until experimentation became an accepted part of physics, beginning in the middle of the sixteenth century, that significant progress was made in science. The result was a dramatic shift in the practice of physics and the introduction of the Newtonian

---

*A Planck length, named after German physicist Max Planck, is 1.6 x 10-35 metres.

world view. Isaac Newton combined philosophical theory and experiment, and where the necessary mathematics was missing, he invented it.* Thus was the triumvirate of theory, mathematics and experimentation established and put to good use for the next three centuries.

Mathematics is now so widely accepted as the arbiter of truth in the modern world that it has become the backbone of disciplines ranging from physics (of course) to economics and sociology. Backing up a statement with mathematics gives it an aura of validity, even if the topic has to do with something as mathematically messy as human behaviour. The popular CBS television series *Numb3rs* provides a good illustration of the Western world's belief in the power of mathematics. In each episode, two mathematical geniuses are brought in by the FBI to solve intractable cases. By using the magical power of complicated mathematics, they can predict everything from the location of a hostage to where a sniper will strike next. Like modern dragon slayers, the mathematicians always save the day . . . and the hostage.

Of course, *Numb3rs* is theatre. In the real world, predicting human behaviour is enormously complicated. Just ask the people who utilize math programs in an attempt to predict how people will vote or the next big thing on the stock market. Reading chicken entrails or going with a gut feeling can often be just as effective as attempting to crowbar human behaviour into a mathematical model.

The popularity of mathematics as a truth-teller lies primarily in the certainty it provides. Mathematics comes with proofs. A mathematician first devises a set of rules or axioms, and then uses a series of logical statements to arrive at a conclusion. If there are no flaws in the logic, the conclusion is proved. That's it. No messy testing is required.

---

*Isaac Newton and Gottfried Leibniz are both credited with coming up with calculus in 1684, working independently. But Eastern mathematicians had been developing elements of calculus long before it became part of Western science.

There is, of course, nothing wrong with using logic as a means of arriving at an understanding of the natural world. But logic has its limits. In the sixth century BCE, the people of Elea in southern Italy were caught up in the intense excitement of logic as a new and novel method of reasoning. "Logic was the rage of the day; all over, in the marketplaces, in the streets, in private homes and public buildings, at all times, sometimes all through the night, people engaged in dialectic disputations and flocked to hear the acknowledged masters of logical argument display their art." It all got a bit out of hand, however. The appeal of logic as a means of constructing truth was so great that wherever an exercise in logic contradicted obvious physical evidence, the Eleatics happily conceded that logic prevailed and the evidence of their senses was hallucinatory.[6]

The Eleatics are credited with being the first to introduce logic into the search for truth about nature and reality, and they were also the first to demonstrate, in a classic case of *reductio ad absurdum*, the silliness that can follow when logic alone is the arbiter of truth.

Where the Eleatics would say logic is truth, modern physicists would say mathematical logic is truth. Indeed, no idea in modern physics can be considered credible until it is validated by mathematical logic. It would make no sense to float the idea of dropping experimentation as a means of determining truth unless you have profound faith in the power of mathematics. The danger of heading down this path is that modern theoretical physics—the science that explains our most fundamental reality—risks becoming once again a one-trick pony. It didn't work so well for the ancients, given that Aristotelian science stalled for a couple of millennia. Modern theoretical physics has been stalled for only a couple of decades, so maybe it's a bit early to panic. But some theoretical physicists and mathematicians are sounding the alarm bells about the direction theoretical physics is taking.[7]

It may seem quite bizarre that modern physics would seek progress by regressing to a pre-Newtonian form of scientific methodology, but perhaps the signs have been there all along—we just didn't recognize them.

The seeds of the shift currently taking place in science were sown eighty years ago, from 1925 to 1927. That's when a dramatic two-year revolution in physics reached a climax, and the dénouement set the course for what was to follow. It's the story of a rush to formalize quantum physics, the work of just a handful of men fired by ambition, philosophical conflicts and personal agendas.

This was never a team effort. Sometimes, two or three would collaborate for a while, but mostly they were rivals who wanted their particular version of the new science to prevail. They had little enough in common. They were German, Swiss, Austrian, French, Danish and English; from royal blood, academic dynasties and the working class; Lutheran, Jewish, Catholic, Vedantic and atheist. About all they had in common was the fact that they were white, European males. But each of them contributed to the quantum puzzle a crucial piece needed to formalize the new science, and each played a role in a quantum revolution that stalled in a pressure cooker of tension, tragedy and betrayal.

The development of quantum physics also coincided with the final glory days of a grand scientific dynasty in Germany. The Great War had dealt a heavy blow to German pride, and the humiliation of the Treaty of Versailles, which laid out the terms of German surrender, fuelled the desire to see science return to its former exalted status for the sake of the Fatherland. It also fuelled a rising nationalism that designated Jews and foreigners as scapegoats for Germany's failed expansionist ambitions.

Quantum physics was still very new in the 1920s. Max Planck is typically credited with introducing "quanta" into the lexicon of physics in 1900, and in 1905 Einstein quietly opened a can of worms by using quantization to demonstrate that light could be both a particle and a wave. But it wasn't until Niels Bohr, a young Danish physicist, applied quantization to the planetary model of the atom in 1913 that anyone was much interested in this "quantum" stuff.

It was a tricky start. Bohr's atomic model introduced the strange idea that electrons which orbited the nucleus of an atom could jump from one allowed orbit to another without traversing the space in between. It's like standing on one side of the room and then suddenly being on the other side, without ever crossing the floor. That type of behaviour is not allowed in the classical world, but it's what seemed to be happening in the atomic world. Some physicists were so offended by this violation of classical physics that they threatened to quit the profession if "quantum jumps" turned out to be true.

It quickly became apparent that the quantum world of atoms and electrons did not seem to be operating by the same rules of physics as the everyday, classical world. In classical physics, energy flowed continuously, but in quantum physics it came in chunks, or quanta. The quantum world allowed that light could be both a defined object and a spread-out motion at the same time, which simply could not happen in the classical world. And then there were quantum jumps to consider. It was getting harder and harder to see how the physics rules could ever be reconciled into a unified picture.

The Great War slowed scientific progress in Europe. Germany had been the undisputed world leader in physics at the onset of the war, and much of the scientific talent was put to work in the cause of the Fatherland, which didn't leave much time for theorizing about the strange new quantum world. By the end of the war Germany was in rough shape, but the science community got back to work rebuilding itself, and there was time to return to the puzzling new science that had just begun emerging in the pre-war years.

By the early 1920s, the quantum revolution was under way. It was obvious that classical physics could not explain the quantum world, but it was far from clear how much of the old science would have to be jettisoned in order to make way for the new. Nor was it clear how the strange impossibilities of the quantum world could ever be integrated into a theory that also included the classical world.

With the increasing reliance on symbolic mathematics to "describe" the quantum world, it was also getting harder and harder to visualize how the microscopic world operated. For some of the quantum revolutionaries, it wasn't necessary to be able to visualize the quantum world as long as their calculations matched experimental results. It didn't matter that the symbols and mathematics they used might not have any link to physical reality; mathematical "truth" was enough. But for others, not being able to create a mental picture of the quantum world, in the same way that Einstein had imagined curved space-time in his classical theory of general relativity, meant the quantum world would be forever out of reach.

This was the problem facing the physicists at the forefront of the revolution. In the years from 1925 to 1927, they were hammering out a mathematically logical physics to describe how the quantum world operated (quantum mechanics) and struggling to come up with a sensible explanation for why the quantum world behaved as it did (quantum theory). They succeeded at one but fumbled the other.

Remarkably, this dramatic shift in science was primarily the work of ten men, and they were ten very fallible men, some famous and some not so famous, although they also had a large supporting cast. Their triumphs and tragedies, loves and betrayals, dreams realized and ambitions thwarted, shaped the competition over who would get to define truth and reality. There never was a consensus. By the time of the pivotal Fifth Solvay Conference in Brussels in 1927, there was so much ill will and disappointment among the creators of quantum physics over their various competing theories and over who deserved credit that most were barely on speaking terms.

The Brussels conference was the first time so many of them had come together: Albert Einstein, the lone wolf; Niels Bohr, the obsessive but gentlemanly father figure; Max Born, the anxious hypochondriac; Werner Heisenberg, the intensely ambitious one; Wolfgang Pauli, the sharp-tongued critic with a dark side; Paul

Dirac, the quiet one; Erwin Schrödinger, the enthusiastic woman-izer; Prince Louis de Broglie, the French aristocrat; and Paul Ehrenfest, who was witness to it all. Their coming together, how-ever, lasted only for the duration of the conference.

The Solvay conferences, which had begun in 1911, were funded by the wealthy Belgian industrialist Ernest Solvay, and were not at all like modern science conferences. Today's conferences are marathons of multiple simultaneous presentations by scientists with perhaps twenty minutes to speak, typically preceded by harried fumbling to get the PowerPoint presentation up and running, fol-lowed by five minutes of questions from the audience. The Solvay conferences instead focused on a handful of talks by eminent scien-tists, followed by lengthy discussions and debates in which everyone present had the opportunity to speak and be heard. The whole point of the conferences was to thoroughly explore some aspect of cutting-edge physics, giving the specialists in that area a chance to make a case for their preferred theories and the other notables the opportunity to offer, ideally, thoughtful and constructive crit-icism. They were by-invitation-only affairs, gatherings of the who's who of world physics in the particular area under discussion, with meetings taking place every three years (unless there was a world war going on).

The attendees of the Fifth Solvay Conference boasted seven Nobel prizes in physics (eight if two-time winner Marie Curie's Nobel for chemistry is included), and of the twenty-nine scientists in attendance, another nine would later receive Nobel awards. The future laureates included Heisenberg, Pauli, Born, Dirac, de Broglie and Schrödinger, joining Einstein, Planck and Bohr in winning this most prestigious physics award for their contributions to the creation of quantum physics. Pascual Jordan, who worked with Heisenberg and Born, was not invited to the conference, per-haps because his pronounced stutter made public speaking a trial for both him and his audience. Jordan was also passed over for a

Nobel Prize, despite numerous nominations, but that may have had much more to do with his ardent pro-Aryan leanings.

Immediately after the Brussels conference, personal tragedies, politics and war overtook the fragile new physics while it was still tottering on shaky legs. The men who had created quantum physics scattered, and there was no going back to ponder the problems that had been left unresolved before the rush of experimental success in the 1930s made looking back seem quite unnecessary.

By their very nature, revolutions are fraught with uncertainty about the future, but the breaking apart of old structures and belief systems seems necessary in order to allow the light of new ideas to penetrate. There was no obvious point at which the old order in physics was overthrown by the new quantum variety—no queen beheaded, university razed or tyrant whisked off to a country without extradition treaties—but neither was there a specific point at which Newtonian physics replaced Aristotelian physics. Yet if one were pressed to identify a crucial turning point in the quantum revolution, the 1927 Solvay Conference would be a good choice. The conference loomed as a deadline for the physicists who would present to the august ensemble three competing versions, even if they were hastily-hammered-together and incomplete presentations; it was also the catalyst that finally brought them together in the same place.

Like the showdown between Wyatt Earp and the Clanton gang at the OK Corral, the conference was a fight to see which side would emerge the winner. Unlike the shootout at the famed corral, however, when the shouting was over in Brussels, there was no nice, tidy conclusion with clearly defined winners and losers. The story of the "quantum ten" and the confusion and uncertainty that have bedevilled the development of quantum physics undermine the idea that quantum physics was all figured out a long time ago.

After the upheaval of the Second World War, the story of the development of quantum physics was rewritten and a scientific mythology emerged. According to this postwar revision, the version of quantum physics promoted by those working with Niels Bohr in Copenhagen had won the day at the 1927 Solvay Conference and been declared the winner. And, this story insisted, all the problems with quantum physics had been settled at that time. There was, therefore, no need to revisit the creation of quantum physics because all the issues and questions had already been dealt with.

This hardly makes sense when leading physicists have been saying for years that nobody understands quantum physics. Nobel laureate Murray Gell-Mann is often quoted as saying, "We all know how to use and how to apply it to problems; and so we have learned to live with the fact that nobody can understand it."[8] Richard Feynman said much the same in 1965 in a BBC interview, and so did Steven Weinberg when he admitted in 1992 that he was a little uncomfortable knowing he'd spent his life working within a field that nobody understands.[9] How could all the questions about quantum physics have been answered if it's still not understood by its practitioners?

There continues to be a lot of resistance in the physics community to talking about such questions, since it does rather undermine the authority of the professionals to whom people look for an explanation of their world if the professionals don't understand it themselves. In its own way, this fundamental incomprehensibility of quantum physics has become the proverbial elephant in the living room.

Physicists and their students are still hearing the argument that quantum physics was fully worked out and satisfactorily explained eight decades ago, and popular science books reinforce that idea by routinely replaying the postwar mythology of the development of quantum physics. Indeed, as science writer George Johnson recently noted, "The story of the quantum revolution has been told so many times that it has become as ritualized as the stations of the cross."[10]* But another look at the turbulent development of the new science, including its very fallible creators, tells a much different tale of what really happened—a story of passion, tragedy, ambition and science. Perhaps a re-examination of how quantum physics was created can shed some small light on why, after eighty years, physicists are still having so much trouble reconciling the classical and quantum worlds. And preferably, this reappraisal will take place before modern science starts heading down the path taken by the Eleatics.

---

*The stations of the cross, part of a ritual of the Roman Catholic church, are depictions of fourteen events in the journey of Jesus from the time he was condemned to death until he was laid in his tomb. The faithful visit each station in turn, meditating and praying.

Chapter Two

THE QUANTUM SHOWDOWN

Our life is divided betwixt folly and prudence: whoever will write of it only what is reverent and canonical will leave one-half behind.

MICHEL DE MONTAIGNE

The art deco salon, with its stylish mosaics, opulent wood pan-elling and stained glass windows, was a suitably grand backdrop for what was shaping up to be a dramatic and pivotal moment in the development of modern science. The Fifth Solvay Conference at the Institut de Physiologie in Parc Léopold, Brussels, began October 24, 1927, and there was a powerful sense amongst the assembled scientists that they were witnessing a major turning point in the world of science.

Right from the beginning, there was a palpable excitement in the air. Almost all the key players in the development of quantum physics were there. The theme of the conference was "Electrons and Photons," but everyone knew it was really about quantum physics. Over the previous couple of years, an incredible amount of innovative thinking and sheer hard work had gone into formulating the new science, but there was no consensus about how the new physics worked or what it meant. And now the handful of scientists who had hammered it all together were about to present to their peers three competing versions—three factions hoping to persuade the august assembly that their theory was the right one.

Paul Ehrenfest had a niggling feeling that he was at the conference under false pretenses. He was, in his own estimation, an underachieving scientist from a small university in the Dutch city of Leiden. He wasn't sure he belonged in this rarefied company of Nobel laureates and the mathematically gifted young scientists who had done so much to make quantum physics a real science. He'd once revealed to his dear friend Albert Einstein, "I hop around among all of you big beasts like a harmless and helpless frog who is afraid of being squashed."[1] He still felt that way.

This wasn't Ehrenfest's first Solvay appearance. He'd been at the third conference, in 1921, but that time he'd presented a paper on behalf of Niels Bohr, another of his dear friends. The focus of the 1921 conference had been "Atoms and Electrons," and it would hardly have been a thorough discussion without a contribution from Bohr about his remarkable, if baffling, model of the quantized atom. But Bohr had been under a lot of stress at the time, overseeing the opening of his new Institute of Theoretical Physics in Copenhagen, and together with the demands of his expanding family, he'd worked himself into a state of exhaustion. Ehrenfest stepped in and offered to read Bohr's not-quite-finished paper, along with work of his own, and save him the trip from Copenhagen to Brussels.

Ehrenfest knew just about everybody at the conference, some very well. A quick glance around the room revealed half a dozen Nobel laureates, Bohr and Einstein among them. The white-bearded Hendrik Lorentz, whom Ehrenfest considered a father figure, had a Nobel Prize, as did the recently retired doyen of German physics, Max Planck. Marie Curie was still the lone woman at Solvay. She had a Nobel Prize for physics, likewise British experimentalist Lawrence Bragg. American experimentalist Arthur Compton had just got the nod for the 1927 prize. If Ehrenfest felt just a little inferior, he figured he had good reason.

But Ehrenfest was intimidated by more than just the award winners. In terms of career advancement, it was a significant honour to be invited to a Solvay conference, and it was an even bigger honour for a small group of young men with the ink barely dry on their doctoral degrees. They formed a kind of "boys' club." Paul Dirac was the youngest at twenty-five, Werner Heisenberg was almost twenty-six and Wolfgang Pauli, twenty-seven. Pascual Jordan hadn't been invited, even though it could be argued that his mathematical contributions were on par with those of Dirac. Jordan had a terrible stutter that ruled out public speaking, but, given that he had yet to complete the habilitation that was required after receiving a doctoral degree, he was unlikely to have rated an invitation to such an elite gathering anyway. Jordan was born a few months after Dirac, so if he had been invited, he would have been the youngest scientist in the room. At times like this, Ehrenfest could feel every one of his forty-seven years.

With one exception, the boys' club was collected, as usual, around Bohr. Heisenberg, Pauli and Dirac had all spent time with Bohr in Copenhagen, where Heisenberg had also been his assistant. All three had also been at Göttingen University over the past two years, working with Max Born, another of Ehrenfest's friends. Pauli and Heisenberg had both served as Born's assistants, although Pauli had lasted only a few months in the quiet backwater of Göttingen before heading to the big city lights (and nightclubs) of Hamburg. Jordan was Born's current assistant. To the physics community, Bohr, Heisenberg, Pauli, Dirac, Born and Jordan were the key players in the formidable Copenhagen–Göttingen school of quantum physics.

Born and Ehrenfest had been friends for a long time, so Ehrenfest knew what was going on at Göttingen. And because Ehrenfest and Bohr were also close friends, he knew what was happening at Copenhagen. For most of the preceding two years, the collaboration on the quantum atom that had been going on between Copenhagen and Göttingen seemed to have worked rather well. Heisenberg, Pauli and Dirac travelled back and forth between the two institutions, all of them spending time with both Bohr in

Copenhagen and Born at Göttingen. Only those who were privy to what was really going on in the two research centres knew about the friction and personal animosities playing out behind the scenes.

To the rest of the physics community, the Copenhagen and Göttingen collaborators presented a facade of unity and harmony. They knew it made their version of the new physics stronger if it looked as if they were all singing from the same songbook. But by the time of the conference, no one knew better than Born how much the Göttingen efforts were being sidelined. The new physics was being presented under the Copenhagen banner, as if the Göttingen contribution had been trivial or inconsequential. It wasn't Copenhagen and Göttingen anymore; it was just Copenhagen. It was obvious to Born that Heisenberg was shouldering both him and Jordan out of the picture, but it didn't seem that Born could summon the energy any longer to fight for Göttingen's fair share of the credit.

Heisenberg wasn't orbiting around Bohr like some of the other young men. He'd been a Bohr protégé, but that was over now. Heisenberg was the quintessential handsome German lad, fair-haired and blue-eyed. Despite his youth, he had buckets of self-confidence, as well as a cheeky grin that could disarm his colleagues even as he was "borrowing" their ideas. While Pauli and Jordan didn't seem to mind, Born did.

Ehrenfest knew that Bohr and Heisenberg had been at logger-heads during the past year, clashing over how to create a coherent theory of the new quantum physics. It had been a painful time for Bohr, and the two men could no longer work together. But word was out that the ambitious Heisenberg had scored the kind of coup he'd been shooting for. His self-serving competitiveness had just paid off. Shortly before the conference, Heisenberg was notified that he'd been selected as a professor of theoretical physics at the University of Leipzig, and at a lucrative salary. He was about to become Germany's youngest full professor, with every intention of establishing Leipzig as the world centre for the study of the new quantum physics.

Pauli, swaying back and forth as he often did, looked glum. Was he hung over? The portly Pauli was very fond of the nightclub scene and had something of a reputation for his copious alcohol consumption. It wasn't something Ehrenfest had witnessed first-hand, even when Pauli visited him in Leiden. Ehrenfest himself was an abstainer. Smoky, noisy bars were not his cup of tea, and his wife did not allow alcohol in the house. The Hamburg nightclubs were known to be Pauli's refuge when he was having emotional difficulties, and even when he was not. Pauli had once told a friend, "I've found drinking wine to be very agreeable. After the second bottle of wine or champagne, I usually become good company (which I never am when sober) and can, under these circumstances, make a very good impression on my surroundings, especially if they are feminine!"[2]

Pauli demanded a great deal of himself, pushing himself to make significant breakthroughs in physics and sinking into depression whenever he hit a dry spell. Pauli also had a reputation for being brutally critical of his Hamburg students and physics colleagues should he spot a shaky argument or weak assumption in their work, which earned him the nicknames "whip of God" and "frightful Pauli."[3] As long as one wasn't the target of his sharp tongue, Pauli's appreciation for a clever prank and his endearing clumsiness made him easy to like. Ehrenfest sometimes teasingly addressed Pauli in his letters as "Sanct Pauli," a reference to the red-light district and nightclub scene that Pauli frequented.[4]

Ehrenfest knew that Pauli and Heisenberg had been students together at the University of Munich, and that Pauli was one of the few people whose criticism the rather arrogant Heisenberg would listen to. When Heisenberg started galloping off in some new direction with a poorly worked out theory, it was usually Pauli who reined him in. But things had grown a little tense between the two over the past few months. When Heisenberg turned to Pauli for help when he and Bohr were fighting over the final details of the theory for quantum physics, Pauli had been unusually distant and unsupportive. In the stressful weeks before the Solvay conference, Bohr ended up looking to Pauli for help, not Heisenberg.

Ehrenfest didn't know a lot about Dirac, the youngest of the boys' club at the conference. Dirac was tall and thin and already a little stooped. Dressed in a sombre black suit, he looked for all the world like an anxious undertaker. The taciturn young Englishman seemed to have appeared out of nowhere two years previously, armed with his own take on the mathematics of quantum mechanics. He was definitely not the chatty sort; someone had jokingly suggested that his vocabulary consisted solely of "yes," "no" and "I don't know." Dirac preferred to let his mathematics do the talking for him. He had been to Leiden for a few days that summer, but even Ehrenfest, with his keen interest in people, had been unable to get the introverted Dirac to open up.

Ehrenfest watched Max Born chatting quietly with Marie Curie and Paul Langevin, and what he saw worried him. He wasn't concerned that Born's attentions to the only woman in the room were inappropriate—far from it. Born was very attached to his wife, and Curie's days as a femme fatale were long past. She was now a sad-faced and grey woman approaching her sixtieth birthday, her eyes clouded with cataracts. It was not easy to imagine this severe and unsmiling woman being at the heart of a romantic scandal. It had started in 1911 when Curie and Langevin, colleagues from Paris, both attended the First Solvay Conference. They arrived home from Brussels to face public outrage because Langevin's vengeful wife had gone to the papers with the story that her husband had eloped with Curie.

The romance between the widow Curie and the married Langevin caused such a public outcry that the Swedish Academy, about to announce Curie's second Nobel award, was having second thoughts. The Academy went ahead with the award but suggested to Curie that she not attend the ceremony, a piece of advice she ignored. The scandal spawned at least five duels in defence of Curie's honour. Fortunately, the duels resulted in only minor wounds, and the challenge Langevin took up resulted in no bloodshed at all and left his trademark handlebar moustache intact.[5] Sixteen years later, the grand passions of that affair had long since cooled.

Einstein was at the 1911 Solvay Conference, and had been amused by the idea of Curie being at the centre of a romantic scandal. "She has a sparkling intelligence," he said of her, "but in spite of her passionateness she is not attractive enough to become dangerous to anyone."[6] Ehrenfest knew all too well that Einstein's taste in female companions ran to something rather more flamboyant than Madame Curie—or Einstein's own wife, for that matter.

Born, on the other hand, was very much in love with his wife, Hedi. Ehrenfest and Born had been friends for a long time, and Ehrenfest knew that his friend tended to fuss over his health, often retreating to a spa or resort to recover, and right now he didn't look at all well. Part of his problem was the travelling. In the last few weeks Born had been to Italy for a conference and then headed to Bristol, England, to receive an honorary degree before rushing to Brussels. He was spending a great deal of time away from home and Hedi. Born loved her desperately, but there was a chill in their relationship that was hard to miss, and he was afraid he was losing her.

Born's worry was justified. Hedi had confided to Ehrenfest that she had befriended a mathematician who lived just down the street from the Born household, a man who just happened to be one of Ehrenfest's long-time friends from their Vienna school days. The conference was hardly the time or place to break the news to Born. He already looked like a man on the edge of a nervous breakdown.

Rather, Ehrenfest would have liked to ask Born if he was concerned about his current assistant at Göttingen, Pascual Jordan. There was no question that Jordan was a brilliant young mathematician, but he was also an ardent right-wing nationalist. Jordan was certainly not spouting nasty anti-Semitic propaganda like some physicists at German universities, but maybe that was only because

Jordan's speech impediment made it difficult for him to complete a sentence. Even though Born had converted to Lutheranism to please his wife, he was still a Jew. So were Ehrenfest and Einstein, and even Bohr was half Jewish. Jordan, as far as Ehrenfest knew, had always treated Born with respect, but Göttingen, like other German universities, was proving to be an attractive recruiting ground for the fervour and fanaticism of the National Socialist Party headed by Adolf Hitler.

Ehrenfest wondered if Born tolerated Jordan's politics out of plain old-fashioned guilt. Jordan had written a paper on a new aspect of quantum physics nearly two years earlier and had given it to Born for publication in a physics journal. Unfortunately, Born had packed it away as he headed off on yet another trip, and he didn't find it until six months later. By then, both Dirac and an Italian scientist had come up with the same idea and sent their work off for publication, while Jordan's paper lay at the bottom of one of Born's suitcases. Born still felt sick about the damage he feared he'd done to the young man's career.

Ehrenfest also knew that Born was still hurt by the souring of his long friendship with Einstein. They were both scientists; they should have been able to get past their ideological differences about quantum physics. In Born's opinion, Einstein had turned his back on the struggling Copenhagen and Göttingen physicists right when they were lost in a quantum wilderness and just when the "true believers," as they considered themselves, needed their leader and standard bearer.

Even without Einstein's help, the Copenhagen and Göttingen schools had, it appeared, finalized their version of quantum mechanics, but there was competition. Erwin Schrödinger and Prince Louis de Broglie both had their own versions to sell, but neither belonged to a "school" of physics. They, like Einstein, worked alone.

Ehrenfest recalled his first meeting with Schrödinger, back in 1912 at the Boltzmann library in the University of Vienna, his old alma mater. He'd dragged the young man, who had only recently graduated, off to lunch and then to a coffee house where he knew they could write equations in pencil on the white marble tabletops while enthusiastically discussing the latest developments in physics.[7]

Schrödinger was the most obviously unconventional person in the conference room. Perhaps it was something he deliberately cultivated. He had arrived in Brussels dressed in lederhosen, with a rucksack on his back, as if he'd just come down from an Alpine hike.[8] Right after the conference Schrödinger would be off to the University of Berlin to begin his duties as Planck's successor as chair of theoretical physics. Ehrenfest wondered if Berlin society was ready for the Viennese scientist.

Schrödinger was quite unabashed about his enjoyment of the pleasures of the flesh, particularly if the flesh belonged to nubile young women. Beak-nosed and bespectacled, he wasn't exactly the image of a Casanova, but he and his wife, Anny, had been part of a rather liberal crowd in Zurich. It was not only acceptable for husbands and wives in their circle to have extramarital affairs, it was expected.[9] Ehrenfest had heard that Schrödinger's big break-through in quantum mechanics two years earlier had come during a holiday tryst in the Alps with one of his many lovers. Ehrenfest couldn't quite fathom the Schrödinger marriage. He adored his own wife, Tatyana, even if they were going through a bit of a bad patch at the moment, and he simply could not imagine himself getting involved with another woman. Still, Berlin nightlife had become a carnival of frenzied licentiousness and excessive passions. Maybe the Schrödingers would fit right in.

But right now, Schrödinger had a fight on his hands. He was about to square off against the Copenhagen and Göttingen physicists in an attempt to convince the assembled scientists that his version of quantum physics was much better than theirs. The theoretical physics that Schrödinger had come up with on his erotic Christmas holiday had stunned the Copenhagen and Göttingen crews, especially since they had assumed they were the only game in town. How had an outsider, working alone, come up with a different explanation for the quantized atom that looked so much better than theirs? It wasn't long before the Copenhagen and Göttingen scientists were integrating Schrödinger's mathematics and physics into their own version, even as they publicly ridiculed his work. Things had got a little out of hand for a while when Heisenberg and Jordan went overboard with their attack on Schrödinger. But that was only because Schrödinger's version of the quantum atom was immediately seen as a serious challenge to Copenhagen's primacy.

Since Schrödinger's mathematics was now integrated into the Copenhagen model, which was far and away the preferred method being used by physicists, he'd already won half the battle. But could Schrödinger persuade the audience that his interpretation of quantum physics should also prevail? The Copenhagen group had come up with their interpretation only in the last couple of weeks, making sure it was obviously and clearly different from what Schrödinger was going to pitch. But it was a confusing theory. Fortunately, the Copenhagen group had the credibility of the greatly respected Niels Bohr on their side, as well as the powerhouse of the two leading theoretical physics institutes, Copenhagen and Göttingen. On his side, Schrödinger had . . . well, just Schrödinger. And maybe Einstein and Planck.

Although most of the people in the room were invited scientists, some were younger men who had been chosen to act as scientific secretaries, taking notes on the proceedings and writing them up for publication after the conference. Even if they were not formal participants, the scientific secretaries had the rare opportunity to rub shoulders with some of the top physicists in the world, and the privilege of a ringside seat at a most important meeting. Maurice de Broglie had been one such fortunate beginner. He'd completed his doctoral degree in 1908 under the supervision of Paul Langevin in Paris, after defying his family's wishes and trading a naval career for physics. He'd barely begun his studies when his father died and Maurice inherited the title of sixth duc de Broglie, and not long afterwards, he'd been invited to accompany Langevin and Curie to the First Solvay Conference in 1911 as a secretary.

Ehrenfest remembered meeting Maurice at the Solvay conference of 1921, where Maurice was a presenter as well as a scientific secretary. His family's wealth and prominent social status in France did not exempt him from having to make his mark in the physics community the same as every other scientist. Maurice had built his own laboratory in his large house in Paris, and he had made a name for himself as an excellent experimentalist. Apparently, Maurice had brought his younger brother Louis with him to the First Solvay Conference back in 1911, and that had inspired Louis to become a physicist himself. Now, Prince Louis de Broglie was about to present his own theory of quantum physics to the assembled scientists. It was too bad, thought Ehrenfest, that Maurice couldn't be present to see it.

Ehrenfest didn't know the young prince. The de Broglie family occupied a social stratum that was very different from anything Ehrenfest had ever experienced. They had a long history in politics and diplomacy in the upper echelons of French society, and at least one de Broglie had lost his head to Madame Guillotine in the French Revolution. Ehrenfest, on the other hand, was the son of a Viennese grocer, so Parisian royal affairs were foreign to the world he knew.

But Ehrenfest did know that Einstein had initially supported Louis de Broglie's version of quantum physics and then backed off a bit. That signalled to Ehrenfest that de Broglie was likely to have a tough time bringing the assembled scientists onside. Certainly, the young prince had done little to endear himself to the Copenhagen scientists. If anything, he seemed to have gone out of his way to irritate them by challenging the science they were doing. It didn't help de Broglie's case that his own ideas were often wrong, or that, with his crooked teeth, *petite moustache* and mop of curly hair, he looked more like the popular cinema comedian Charlie Chaplin than a serious scientist. He would be a sitting duck for Heisenberg's ridicule and Pauli's acid tongue.

At any rate, the assembled scientists and secretaries quickly found their chairs as Lorentz took his place at the front of the salon. Lorentz had chaired all four of the previous Solvay conferences as well. The retired Dutch scientist was a measured and diplomatic man with more than the necessary fluency in German, French and English. But most important, he possessed the ability to keep the sometimes opinionated and headstrong scientists in order. The beautiful salons of the Institut de Physiologie were hardly suited to the rowdy, unconstrained behaviour of a beer hall. Not that Ehrenfest would have minded that, sans the beer, of course. He quite enjoyed an enthusiastic debate, but his colleagues often seemed more at home in a formal atmosphere that he felt worked against an open exchange of ideas. Still, there was a buzz of excitement in the room that all the starched collars and tightly buttoned waistcoats could not suppress.

Adding to the undercurrents was an awareness that this was the first time German scientists had been allowed into Belgium since the end of the Great War. The last time Germans had entered Belgium, things had not gone at all well. German scientists had been banned from attending all international science conferences after the surrender in 1918. Einstein was an exception, partly because he had long ago obtained Swiss citizenship, partly because he had become an international celebrity. But since German scientists had done so much of the work on quantum physics, it would have been pointless to hold a conference on the new science without them.

Lorentz, who also organized the conference, had sought a personal audience with Albert I, King of the Belgians, to plead on behalf of the German physics community. According to Lorentz, he and the king had seen eye to eye on the issue of anti-German sentiment:

His Majesty expressed the opinion that seven years after the war, the feelings which they aroused should be gradually damped down, that a better understanding between peoples was absolutely necessary for the future, and that science could help to bring this about. He also felt it necessary to stress that in view of all that the Germans had done for physics, it would be very difficult to pass them over.[10]

Still, the echoes of the war were an unwelcome reminder of the troubling world beyond the scientific presentations, which all the professional camaraderie couldn't hide. Everyone at the conference knew about the vitriolic attacks Einstein was subjected to in Germany. There were rumours of assassination attempts, or at least death threats. For some Germans, the surrender of 1918 had been both humiliating and enraging. How could they have been defeated when not a single battle had taken place on German soil? So they looked for scapegoats to blame for Germany's defeat, and the Jews were a convenient target.

The success of Einstein's theory of general relativity, experimentally verified just as the war ended, had made him an international star. It had also provided the right-wing factions in German politics with a highly visible target. With armed militias roaming the streets of Berlin, inciting violence and chaos in order to undermine the shaky republican Weimar government, it seemed almost inevitable that highly placed Jews would be targets for German wrath. In the summer of 1922 the foreign minister, Walther Rathenau, Germany's highest-placed Jew and one of Einstein's friends, was assassinated in the street. More and more political figures, Jewish or otherwise, were falling victim to right-wing nationalist paramilitary forces, and the German government seemed unable to put a stop to the violence. Max Planck, one of Germany's leading physicists, who had helped bring Einstein to Berlin in the first place, had pleaded with him not to abandon Germany. But in a quiet talk with Planck, Einstein acknowledged the jeopardy he faced. "A number of people who deserve to be taken seriously," Einstein had told him, "have independently warned me not to stay in Berlin for the time being, and, most especially, to avoid all public appearances in Germany. I am said to be among those whom the nationalists have marked for assassination."[11]

Ehrenfest and his wife did their best to persuade Einstein to come and stay with them in Leiden in 1922, but just as during the Great War, Einstein had declined to leave Germany unless the situation became dire. Fortunately, he had already committed himself to lecturing overseas in the fall of 1922, and thus had a good reason to leave Berlin. He boarded a ship sailing for Japan in October, thereby missing the announcement that he'd won a Nobel Prize for physics. He missed the ceremony in December as well.

Despite the political instability and unrest in Germany, the big question buzzing around the conference was, "What will Einstein say?" Indeed, "all over Europe"[12] people were waiting to hear Einstein's verdict on the new physics. Because of his celebrity status around the world, just about anything Einstein said made big news. But he had yet to make a public statement about quantum physics, so no one could know for sure what he was going to say, not even the physicists gathered in the salon.

Nobody was more aware than Ehrenfest of the deep divisions between Einstein and Bohr over the direction the new physics was taking. But what Einstein would say at the conference, to be recorded as part of the published proceedings, was not necessarily what he would say in the privacy of Ehrenfest's living room. Ehrenfest had hoped Bohr and Einstein could find some common ground, maybe compromise a little. He'd already tried to mediate by arranging some rare uninterrupted time for his friends to sit down and talk to each other, but his best efforts had failed.

Bohr wasn't giving a presentation at the conference, but the interpretation of quantum physics he'd presented a month earlier at a conference in Italy was undoubtedly going to come up for discussion. Ehrenfest hadn't been there, but some of the others at the conference had. Ehrenfest just hoped he could follow what Bohr was trying to say; he'd learned long ago that understanding Bohr wasn't easy. When Ehrenfest had visited the Copenhagen Institute for the first time in 1921, Bohr's younger brother Harald, a professor of mathematics in Copenhagen, had reassured Ehrenfest that he wasn't alone. "When Niels tells me something," Harald confided to Ehrenfest, "I absolutely don't understand what he is talking about and what he is driving at for fifty-nine minutes; but in the sixtieth minute a light suddenly dawns and I see that everything he said previously was absolutely necessary."[13]

Einstein had not been at the conference in Italy either, so no one really knew what his reaction would be to Bohr's new theory. Bohr had sent Einstein a copy of his talk, but this would be the first time the two men would have an opportunity to discuss it. Ehrenfest had a sinking feeling that his two dearest friends were heading for a showdown, a clash of the titans of science—and only one would come out a winner.

Chapter   Three

THE   BIRTH   OF   THE   QUANTUM

Very strange people, physicists—in my experience the
ones who aren't dead are in some way very ill.
> MR. STANDISH in *The Long Dark Tea-Time*
> *of the Soul* by Douglas Adams

Paul Ehrenfest, more so than his friends Einstein and Bohr, was a direct product of the German and Austrian scientific dynasty that was the powerhouse behind the study of thermodynamics, and which had spawned the new physics. It was Ludwig Boltzmann, Ehrenfest's doctoral supervisor, who had set the stage for quantum physics.

During the late 1800s and early 1900s, physicists were puzzling over the fact that some materials absorb and emit heat while others, such as the filaments in electric light bulbs, emit light. The investigation was particularly intense in Germany, where scientists— as civil servants—were expected to work closely with industry to enhance the image of the Fatherland as the world leader in invention and industry.

Germany had been a little slow getting onto the industrial bandwagon. The German state was established in 1871, consolidating various competing factions and dukedoms under a single banner. The creation of the German Reich brought together Prussia's military might, a population larger than that of any other European power, and lots of natural resources to be exploited under an authoritarian and militaristic regime. Under Kaiser Wilhelm II and the firm hand of Imperial Chancellor Otto von Bismarck (the Iron Chancellor), the new Germany was in the ascendant.

Bismarck laid the groundwork for success with his recognition that military might must be accompanied by industrial growth. Industry incorporated research and development into its business practices, providing a fertile arena for scientists and technicians. Before the Great War, no other country could touch Germany's achievements in science, industry and the arts. The Gründerzeit (Founder's Era) at the turn of the twentieth century was a catalogue of industrialists and inventors whose names are still recognized: automobile inventors Carl Benz and Gottfried Daimler; diesel engine inventor Rudolf Diesel; electrical appliance manufacturer Werner von Siemens; airship inventor Ferdinand von Zeppelin; and steel manufacturers Friedrich Krupp, Leonhard Höesch and August Thyssen, to name but a few.

The European cities of Berlin, Prague and Vienna were close enough to each other to permit and even encourage the cross-fertilization of ideas. This cultural and intellectual hotbed produced a wealth of artists and thinkers: composers Johannes Brahms and Richard Strauss; philosophers Friedrich Nietzsche, Karl Marx and Ludwig Wittgenstein; newsmen Joseph Pulitzer and Paul Julius Baron von Reuter; authors Thomas Mann and Franz Kafka, and poet Rainer Maria Rilke; artists Paul Klee, Max Ernst and Gustav Klimt; psychoanalyst Sigmund Freud and neuropathologist Alois Alzheimer; and designer Walter Gropius.

Universities in Germany were world leaders in research, particularly in science. The relative ease of travelling between major European universities by train or boat allowed researchers to collaborate, or just keep in touch. A corollary of this situation was that scientists working in universities outside Europe and England were out of the loop.

Germany at the turn of the twentieth century was definitely the place to be for aspiring scientists. Indeed, German was the language of physics. It's not that scientists weren't doing some good work at other universities in such places as England, France and Austria, but ambitious young scientists headed for Germany to spend a year or two at the universities of Göttingen, Berlin and Munich, enhancing their career opportunities by studying with the great men of German science.

In 1900, almost half the physicists in the world were based in Germany, England, France and the United States. Even so, the United States was pretty much a scientific backwater, with fewer than 100 members in the newly formed American Physical Society.[1] Unlike their German counterparts, American professors were expected to devote their time to teaching, and were not encouraged to do independent research.

In genealogical terms, modern physics in Germany can trace its historical roots back to the establishment of the Berlin-Brandenburg Academy of Sciences in 1700 under the leadership of Gottfried Leibniz. Just about all of today's physicists and mathematicians derive from Leibniz's students of the early 1700s, Leonhard Euler and Joseph Lagrange.[2] Although Isaac Newton was Leibniz's contemporary, competitor and English counterpart based at Cambridge University, Newton did not take on any students. Thus, the British "lineage" ended with his death.*

But the real rise of physics in Germany began with the foundation of Berlin University in 1810. Berlin University begat the Magnus School in Berlin—which was really just a private lab that Heinrich Magnus, the director of the Physical Institute at the university, had built for himself in his house—and out of this school in the mid-1800s emerged a couple of promising young scientists. One of them was a young military surgeon assigned to a regiment of the Hussars near Berlin. When time away from military duties permitted, Hermann Helmholtz worked in the Magnus lab studying

---

*Because Newton had no children—there is no suggestion even of an intimate relationship—he had no direct family descendants either.

thermodynamics—in particular, how matter created heat through
rotting and decay. No one was really sure what heat was: it might
have been a special, invisible, weightless fluid, with hot substances
containing more of this fluid than cold ones; or perhaps heat was a
property of motion, since substances such as metals got hotter and
hotter the more they were rubbed together; or maybe it was the
vibration of the corpuscles of matter.

Helmholtz, at the age of twenty-seven, is credited with coming
up with the first law of thermodynamics, coupled with the law of
conservation of energy.[3] The first law of thermodynamics states
that when heat is added to a system, it must increase the energy of
the system or it must perform some kind of work. The burning
of coal under a boiler, for instance, will generate heat to boil water
to turn a steam turbine to power a generator to provide current to
light an electric lamp. The heat emitted by burning coal must be
transformed into either energy (light emitted by lamps) or work
(the turning of a turbine); hence heat must also be a form of energy.
The law of conservation of energy states that the total energy in a
system—heat, the kinetic energy of motion, the potential energy of
gravity, and chemical energy—must be conserved in time, meaning
that energy can change from one form to another but cannot simply disappear, or appear out of nowhere.

Helmholtz wasn't the only person working on conservation of
energy. James Joule, an Englishman who divided his time between
running the family brewery and conducting energy experiments in
the lab he built in the basement of his parents' home, came up with
the same idea. German scientist Julius Mayer actually came up with
the notion of conservation of energy a few years earlier than either
Helmholtz or Joule, but he was working as a Dutch ship's doctor
in the West Indies and was seriously out of the loop. Mayer, upset
that his work remained unappreciated by the science community,
tried to commit suicide in 1849 by jumping out of a third-storey
window, but he succeeded only in breaking both his legs. He ended
up a cripple in an insane asylum. In a seeming act of contrition, the

German state promptly awarded Mayer the title of "von," the equivalent of a British knighthood.

It's not particularly rare for several scientists to be exploring the same idea at the same time, each unaware of what the others are doing. The person who gets the credit is usually the one who gets published first in a scientific journal, but sometimes the credit goes to the person whose work first draws the attention of the physics community. This time it was Helmholtz who got noticed first, and the first law of thermodynamics and its accompanying law of conservation of energy quickly became the pillars of the study of heat and energy.

The Magnus School also begat Rudolph Clausius, who in 1850, at the age of twenty-eight, came up with the second law of thermodynamics, building on what Helmholtz had started. The second law of thermodynamics states that a system, if left on its own, will inevitably reach a state of equilibrium. In other words, if you set your neglected cup of tea in a pot of cold water, the temperature of the tea in the cup and the water in the pot will eventually be the same. There is no way that your tea will warm up while the pot of water gets colder. This apparent movement from order to disorder Clausius called entropy. When the tea in the cup is warmer than the water in the pot, the system (tea and pot of water) is *defined* as being in a more ordered state—a state that has more rules about how and where molecules can move—than when the tea and water are the same temperature. To put it another way, if all the molecules in the air in a room were to suddenly form a tiny pink elephant sitting in the corner, the molecules would have to have gone from a disordered state to a very ordered one, but the second law of thermodynamics did not allow the process to be reversed in this way.

If the second law of thermodynamics *were* reversible, it would permit, for instance, the impossible pink elephant to suddenly appear in the corner of the room. Or if an egg fell on the floor and broke, it could, in theory, put itself back together again. Reversibility opened the door to all manner of bizarre possibilities that were not observed in the physical world; but irreversibility clashed with the principles of Newtonian physics.

That's where Boltzmann came in. In the 1870s, the bearded young professor with a mop of curly dark hair published a paper that "proved" entropy in the universe always increases.[4] This generated considerable controversy due to the "reversibility paradox." Nobody had a really good explanation for why the second law of thermodynamics should not be reversible when (almost) everything else in classical Newtonian physics was—at least in theory.

Boltzmann too was deeply troubled by the reversibility paradox, and it seemed to him that the only solution must lie in a statistical interpretation. What if the second law of thermodynamics was not a law at all? What if it was indeed reversible but in this chaotic, messy world of ours it's just extremely rare and, hence, highly improbable—but not impossible? Such a solution addressed the paradox, but only by allowing that the behaviour of large numbers of molecules could be described statistically.

At the time, he and other physicists were trying to figure out how to characterize the behaviour of molecules of gas in a container in order to get some insight into the structure and behaviour of molecules. The structure of atoms, if they existed at all, was unknown. But scientists could draw some conclusions about molecules by experiments with a container of gas because it allowed them to vary measurable quantities such as temperature, pressure and volume. Scientists already had a pretty good understanding of the relationship between pressure, volume and temperature, and since those were quantities that could be observed and measured by experiment, they hoped to deduce something about what the gas molecules were doing or how they were interacting. It was the best they could do, given that molecules could be neither seen nor measured.

Boltzmann hit on the idea of using entropy as a defining characteristic of the motion of gases. Scientists working on what was called the kinetic theory of gases assumed that molecules of gas (say, helium) bounced around randomly inside a container at different speeds. The molecules collided with each other and with the sides of the container, giving off heat. At some point, in theory, the gas molecules would all be bouncing around at the same speed (more or less), with each molecule having about the same energy. Once that happened, no more energy would be used to speed up slow-moving molecules or to slow down fast-moving ones, and the temperature throughout the container would be the same. All the energy that was going to be used up inside the container to achieve equilibrium of speed and temperature had been used, and therefore the gas molecules were in a state of maximum entropy. They couldn't get any more disordered.

Entropy wasn't any more measurable than molecules, but it too could be deduced if Boltzmann's probability equation for entropy could be linked to the observable experimental values of temperature, volume and pressure. However, Newtonian physics was just too unwieldy for this problem. Newtonian physics allows, in theory, for each molecule to have a position and a velocity, and all you would need to describe the state of the gas in a container at any point in time is the starting position and momentum (the mass and velocity) of each molecule. Then you could predict, in theory, the position and momentum of each molecule thereafter.

For Boltzmann, this presented an intractable problem. A single gram of gas was assumed to contain about 10^{23} molecules (Avogadro's number). Even if one could know the starting position and momentum of every one of those molecules inside a container, it was simply impossible to do the horrendous amount of mathematics necessary to calculate what was happening to every molecule every fraction of a second. In theory, it was doable; in practice, it was not.

To simplify the problem, Boltzmann figured it was really only necessary to know the average energy, velocity or momentum for the whole ensemble of molecules. That eliminated the necessity of dealing with each one individually. All he needed to do was figure out the most probable state of the molecules, calling the probability W. According to the first law of thermodynamics, if the temperature of a system increased, the energy must increase as well. Thus, Boltzmann could directly link the temperature of the gas to the kinetic energy of the gas molecules by multiplying the temperature by some as yet unknown factor k. But that still didn't describe the relationship between heat and the characteristics of molecules.

According to the second law of thermodynamics, when the temperature inside the container remained constant, the gas molecules would have reached a state of maximum entropy. And that's when the most molecules in the container would have the highest probability of having the average kinetic energy. To calculate the average energy, Boltzmann made use of the mathematics of probabilities, but to make it work, he imposed the condition that energy was divided into chunks, which he knew very well was not really the case. He made it very clear in his 1877 paper that such "chunking" of energy was unrealistic and that he was only using it to help make his methodology easier to understand.

Boltzmann's statistical mechanics offended many in the physics community, particularly in Prague and Vienna. Positivism, spearheaded by Ernst Mach, was the dominant philosophy at the time, and positivism—a philosophy free of religion or other superstitions—was based only on what could be seen, counted or otherwise experienced by the human senses. Mach and his followers roundly condemned Boltzmann's statistical mechanics. Not only did it assume the existence of invisible molecules, it relied on mathematical probabilities—what Mach called a *Lottospiel* or bingo[5]—instead of experimental measurements.

Many physicists, Max Planck included, did not care for Boltzmann's statistical solution to the reversibility paradox. It required accepting a statistical argument to reconcile entropy with Newtonian physics, and that was just not on. Planck, one of Helmholtz's "begats," was an adherent of Machian positivism, albeit of a softer variety, and he had little use for Boltzmann's statistical methods.

Rather than working with gas molecules in a container, as Boltzmann had, Planck was studying thermodynamics using the idea of a container filled with light waves or radiation. He imagined that the walls of the container were made up of tiny spring-like resonators that could expand and contract to absorb and emit different wavelengths. Planck wanted to show that the heat given off by the imaginary resonators inside the container would inevitably reach a state of equilibrium. All the radiation inside the container would be of the same average wavelength, the temperature everywhere inside would be the same . . . and it would stay that way. If he could demonstrate it experimentally, he would have the proof he needed to show that the second law of thermodynamics was irreversible and that Boltzmann's probabilistic model was unnecessary in physics.

Unfortunately, such an experiment was not feasible. However, Planck was able to use what he'd learned about equilibrium states to tackle a particularly difficult unsolved problem in physics: finding an equation that would explain the experimental evidence of the spectrum of visible light from infrared to ultraviolet.

Planck had inadvertently positioned himself at the nexus of three rival paradigms in physics: Newtonian mechanics and the motion of "things," such as atoms and billiard balls; the theory of electrodynamic radiation promulgated by the English physicist James Clerk Maxwell; and the thermodynamics of heat. The physics of particles (things), electromagnetic waves (radiation) and heat had yet to be reconciled into a single theory. The volatile and temperamental Boltzmann, bouncing between the universities in Graz and Vienna, was attempting to reconcile heat with the mechanics of molecules, while the dignified and methodical Planck, at Berlin University, was trying to figure out the relationship between heat and radiation. And both were banking on the energy of an equilibrium state being the same as maximum entropy. Hence, the two men weren't really so far apart in their objectives.

Working closely with experimentalists at the Imperial Institute of Physics and Technology near Berlin in 1900, Planck made a number of attempts to reconcile energy distribution of light with experimental results, and each time he failed. Late in the year, Planck finally turned to Boltzmann's 1877 paper, the one with the "unrealistic" chunking of energy. This was a significant departure for Planck. He had published about forty papers prior to 1900, and not one had made any mention of Boltzmann or the relationship between entropy and the probability of molecular states.[6] But there he was, a member of the school of positivism that opposed the use of atoms and probabilities in science, making use of both atoms and probabilities in his theory of radiation distribution.

Like Boltzmann, Planck proposed that energy was chunked into discrete amounts (quanta), and with that he finally came up with an equation that fit the experimental evidence perfectly. He had introduced a new constant, h, with units of energy seconds,* which he called the "quantum of action."

According to Planck's theory, the "energy elements" of radiation were to be restricted to integer multiples of h, multiplied by the frequency of the radiation. He quickly published his results in December 1900, and followed with two more papers in January, citing Boltzmann's 1877 paper in each. In those two January papers, Planck also derived the numerical value for Boltzmann's k, which Boltzmann had never got around to doing, and formalized Boltzmann's definition of entropy as the logarithm of the probability state multiplied by k.[7]

Planck's derivation for the energy distribution of radiation was readily adopted by the physics community and quickly put to use by the German heating and lighting industries, major funders and beneficiaries of technological research coming out of the institute where Planck's experimental colleagues worked. Few scientists

---

*An "energy second" should not be confused with "energy per second," which is simply force.

seemed worried about *why* Planck's equation worked so well; they used it because it worked, and they remained oblivious to the introduction of quantization and its implications.

But why did Planck make use of Boltzmann's "unrealistic" chunking of energies? Did he overlook the part in the 1877 paper where Boltzmann made it clear that he was using discrete energies only to make his methodology easier to understand? How did Boltzmann's unrealistic discrete energies become Planck's quantized energy elements? Planck himself could not answer these questions.

It is not uncommon in physics to use a placeholder in an equation to make an equation match experimental evidence, and it's done on the understanding that the placeholder is simply a temporary device to be used until scientists can figure out what the placeholder really stands for. But Planck did not treat his h as a mere placeholder; he attributed great significance to his quantum of action. His constant h, Planck said, was part of a "truly natural" system of units, along with the speed of light and the gravitational constant, that would necessarily be identical "for all times and for all cultures, including extraterrestrial and extrahuman ones," and "would retain their meaning so long as the laws of gravitation, of the propagation of light in a vacuum, and the two laws of thermodynamics remain valid."[8]

Planck could not, however, explain the physical significance of his energy elements (h multiplied by radiation frequency). He seemed to be implying that in the microscopic world, energy was chunked into integer multiples of energy elements, yet in the macroscopic world, energy most certainly appeared to be continuous.

It is possible that Boltzmann planted the seed of the quantum (or chunking) in Planck's mind a decade earlier, at a physics conference they both attended. Planck, who was still a staunch Machian at the time, was trying to convince Boltzmann of the superiority of his radiation-based approach over atomism. Boltzmann is said to have announced, "I see no reason why energy shouldn't also be regarded as divided atomically."[9]

Historians of science can work themselves into a fine froth over what Planck did or didn't know when he introduced quantization in his 1900 and 1901 papers. They examine what Planck wrote then and later, searching for clues between the lines to ascertain what he might have meant but did not say, like wizards picking through chicken entrails for insight and enlightenment.[10]

Initially, Planck's quantization was ignored, its significance unrecognized. French physicist and author Étienne Klein later referred to the advent of Planck's constant as the *coup de h* or "axe blow" that destroyed the foundations of the classical world view.[11] (It's a play on words, with *h* phonetically substituted for *hache*, "axe," and is not as funny when you have to explain it.) But at the time, classical physicists continued their work without any concern about Planck's *h*.

Meanwhile, in 1901, Boltzmann abruptly left Vienna for a position at the University of Leipzig. His five years in Vienna had not been happy ones. The philosophical climate of that city was dominated by Ernst Mach, who had moved to the University of Vienna only a year after Boltzmann to become the Professor of Philosophy with Special Emphasis on the History and Theory of Inductive Sciences.[12] They were on cordial-enough terms with each other, but Boltzmann found it difficult to share his realm in Vienna with someone whose philosophy was so openly hostile to his life's work.

Boltzmann had already been diagnosed with neurasthenia,[13] a somewhat general term used at the time to cover general disorders of the nervous system. He was a restless man and found it difficult to stay in one place for very long, hoping a move to another university might ease his anxieties and bouts of depression. Boltzmann was also losing his eyesight and was subject to severe asthma attacks. Before he left for Leipzig, he suffered a complete breakdown and spent the summer in a sanatorium.

When Boltzmann recovered, he completed the move to Leipzig but was already trying to get his old job back in Vienna. Boltzmann's time in Leipzig did not go at all well, and he attempted to kill himself. He was plagued by what one colleague called "the worst sickness that can strike a professor—the fear of lecturing."[14] He was terrified that he would suddenly blank during a lecture.

Unfortunately, Boltzmann's precipitous departure from Vienna had left his student Paul Ehrenfest without a supervisor. It was common practice for university students to travel from university to university, attending the lectures of many leading scientists. Ehrenfest had grown up in Vienna, but not all his years in the city had been happy ones. With his supervisor gone, Ehrenfest had no real reason to stay there, so in 1901 he took himself off to Germany's centre for physics and mathematics, Göttingen University.

Göttingen provided Ehrenfest with a richer menu of studies than Vienna. In particular, he was able to take in mathematics lectures with world-renowned Felix Klein and the up-and-coming David Hilbert. Ehrenfest's interest in mathematics led him directly to a striking mathematics student, Tatyana Alexeyevna. Paul first spotted her in lectures given by Klein and Hilbert, and she was often in the reading room. He started hanging out at the weekly meetings of the math students with the hope of meeting her. When it turned out that Tatyana couldn't attend the math club meetings because women were not allowed in, Paul promptly set about getting that rule changed. He was determined to be introduced. The Russian-born Tatyana and Paul formed an immediate and deep bond when they met, even though Tatyana was four years older, and it wasn't long before Tatyana's rooms became a centre for student meetings and lively gatherings of physics and math students.[15]

In 1903, Ehrenfest concluded his university sampling by spending a few weeks at Leiden University in Holland. In between enjoying the spring tulips and playing tourist, Ehrenfest attended lectures by the well-known physicist Hendrik Lorentz. That's where Ehrenfest was first introduced to Max Planck's work on radiation. In his spare time, Ehrenfest read as many papers as he could by Clausius, Maxwell and Boltzmann, developing a better understanding of Boltzmann's statistical mechanics. He also discovered some new work by Albert Einstein, an unknown from Switzerland. But the highlight of the young man's visit was an evening he spent visiting with Lorentz at his home before he headed back to Vienna.[16]

By this time, the positivist philosopher Mach had had a stroke and retired, and Boltzmann had returned to the University of Vienna. Tatyana and Paul were now engaged, but Paul decided he should return to Vienna to finish his doctoral dissertation with Boltzmann and to continue his extensive study of his mentor's work on statistical mechanics. Tatyana would pursue her studies at Göttingen. They would marry when Paul graduated—and they did, a few days before Christmas 1904, just shy of Paul's twenty-fourth birthday.

Paul and Tatyana Ehrenfest were well suited to each other in temperament, both strong personalities who found a great deal of enjoyment in science and mathematics, and in the animated exchange of ideas. But their backgrounds were very different.

Paul Ehrenfest spent his early years trying to keep out of his parents' way in their busy grocery store in a lower middle-class neighbourhood in Vienna. Their shop and apartment were half of a double house, the other half occupied by a German Catholic family running a butcher shop downstairs and living upstairs, just like the Ehrenfests. The Ehrenfest household was crowded. At one point, the store and apartment housed Paul's parents, his four older brothers (the eldest, Arthur, was seventeen when Paul was born), a nurse-maid for Paul when he was young, two maids, two shopgirls and four male shop assistants.[17]

Young Paul was recognized as being very bright, having more or less taught himself to read and do sums before he started school. He was often the butt of his brothers' pranks, but he was indulged by his grandmother, the undisputed matriarch of the family, and by his many aunts and uncles. Paul's mother died when he was ten, but his father soon married Paul's aunt, his mother's lively and much younger sister, who was about the same age as Arthur. Life continued happily, at least for a while.

By the time Paul's father died, when he was sixteen, Paul had turned into an angry and rebellious teenager whose dark moods bordered at times on suicidal. Arthur, by then an engineer, took over the care of his youngest brother, and it was Arthur's firm hand that kept Paul from going off the rails. Paul had come to despise school, but he stuck with it at Arthur's insistence, and eventually made friends with Gustav Herglotz, a fellow mathematics fan. After finishing school, they were pleased to find themselves at the Vienna and Göttingen universities at the same time.

Tatyana's childhood was quite different from Paul's. She was an only child who travelled by train all over the Russian Empire with her father, the chief engineer of the imperial railways.[18] When he died, she was raised in a middle-class Russian Orthodox household by an uncle and aunt, who had no children of their own.[19] She was certainly not surrounded by family as Paul had been, but Tatyana enjoyed a fairly comfortable life. Her uncle was a professor and encouraged her to pursue her interest in science and mathematics. She completed a university degree at the Women's University in St. Petersburg before looking westward to Germany for the advanced degree that she, as a woman, could not get in Russia. Rather than heading straight to Germany, though, she first spent three years at normal school (teachers' college), studying mathematics and science.[20] That's why she was several years older than Paul when she started her studies at Göttingen University.

When Paul and Tatyana married in 1904, it was still illegal in Germany and Austria for Jews to marry Christians, and since neither wished to change their religion, they elected to marry without a religious affiliation. A few years later this became a serious problem for Paul. Universities in Germany and Austria would not appoint a professor unless he had a formal religious affiliation. Because all professors were civil servants whose appointments had to be approved by the government, they were subject to whatever citizenship and religion restrictions the country placed on government workers. The lack of a religious affiliation also meant that taking a post-wedding trip to St. Petersburg to meet Tatyana's family was a bit tricky as well. The Russian government did not welcome unbelievers. Fortunately, Tatyana was able to persuade Russian officials that she and Paul did not pose a danger to imperial Russia.[21]

Paul continued taking Boltzmann's seminars after the couple returned to Vienna. He and Tatyana both had modest stipends from their family inheritances, so they could afford to stay in Vienna without paid employment. That gave both of them time to pursue their favourite research interests. Tatyana studied the basis of dimensional analysis and Paul studied statistical mechanics, especially Planck's theory on the spectral distribution of radiation. He too spotted the similarity between Boltzmann's model of gas molecules in a container and Planck's model of radiation in a container.

Paul had promised Tatyana that they would move to St. Petersburg if the opportunity arose. In September 1906 they stopped at Göttingen first, just in case there might be a job for Paul there. They were shocked to hear the news that, on September 6, Boltzmann had committed suicide.

After his return to Vienna in 1902, Boltzmann's health had worsened. He'd also agreed to take over Mach's lectures in the philosophy of science, which were initially quite successful, but Boltzmann quickly found that he couldn't keep up with the added workload. His depression worsened, and he stopped teaching by the spring of 1906. On vacation in Italy with his family in September, the day before they were scheduled to return for the fall semester in Vienna, Boltzmann hanged himself from a hotel window casement with a short piece of cord.[22]

Wilhelm Ostwald, one of Boltzmann's colleagues from his time in Leipzig, wrote an obituary for Boltzmann in which he lamented that science demanded so much of its practitioners. "This falls under the laws," wrote Ostwald, "to which almost all servants of the strict goddess Science are subject: their lives end in grief, and this all the more likely, the more completely they have dedicated their lives to Her."[23]

Paul cut the obituary from a newspaper and saved it.

Paul and Tatyana spent several weeks in Göttingen, working together to analyze some of the objections that scientists had raised about Boltzmann's work. Ehrenfest had not been particularly close to Boltzmann, but his death had still come as a shock. Perhaps with Boltzmann's recent passing fresh in his mind, Felix Klein invited Paul to give a talk on statistical mechanics at a mathematics colloquium. Ehrenfest's lecture, based on the work he and his wife had been doing together, excited his audience and impressed Klein.

It just so happened that Klein was in the middle of a massive joint effort by the universities of Göttingen, Leipzig, Munich and Vienna to create an encyclopedia of mathematics. Leading mathematicians and scientists in Europe were being invited to write different sections to provide a comprehensive overview of mathematics and mathematics-based sciences such as physics. Klein, one of the key organizers of the *Mathematical Encyclopedia*, had intended to have Boltzmann write the section on statistical mechanics. Since that was no longer possible, Klein invited the Ehrenfests to write the account of Boltzmann's work. Who better to write the encyclopedia article than one of Boltzmann's last students?

With no job on the horizon in Göttingen, Paul and Tatyana moved to St. Petersburg. Tatyana got a job as a mathematics teacher in a girls' school,[24] and they set to work on the encyclopedia article. According to Tatyana, Paul did most of the research and writing on the project, but it's unlikely he could have done it without her.[25] He needed someone to help him argue through ideas and pick out the flaws in his logic, and Tatyana was an excellent intellectual sparring partner.

Unfortunately, Paul was not happy in Russia. He tried to fit in, growing a beard, learning Russian and letting his hair grow long. He lacked the necessary credentials to teach, but he'd nonetheless been offered a job teaching mathematical physics in 1909. When he was abruptly dismissed a year later, he sank into a paralyzing depression that meant he was doing no work at all.[26]

Fortuitously, Paul's friend Gustav Herglotz wrote that he himself was recovering from alcoholism and depression and invited Paul to join him at a German health spa. While he was there, Paul realized he didn't want to be in St. Petersburg anymore, but that meant he needed to get himself a proper job somewhere other than Russia. Once he returned home, he again sank into a depression.[27]

Paul had been writing to his older brother Hugo, a successful obstetrician in the United States, about his plight. Hugo diagnosed Paul's depression as neurasthenia, the same ailment Boltzmann had suffered from, and advised him that his condition would not improve unless he got out of Russia.[28] Reassured by his brother's support, Paul got back to work and the encyclopedia article on statistical mechanics was finally published in December 1911. And with the birth of his second daughter, it was time for Paul to get serious about finding a job.

Paul sent a copy of the lengthy encyclopedia article, "The Conceptual Foundations of the Statistical Approach in Mechanics," to Hendrik Lorentz in Leiden, who was suitably impressed. Ehrenfest had already garnered a reputation as an engaging lecturer, despite his not yet having acquired a professional position. He had the ability to explain complex ideas in a way that his audience could understand, and he could do it with wit and flair. But Ehrenfest's greatest talent, according to biographer Martin J. Klein, was his ability to critically evaluate other people's ideas:

[Ehrenfest] did not proceed in the manner of most bright young beginners by pushing his master's calculations a step or two further, or by applying his methods to some previously untreated physical problem. Nor did he have the creative genius to propose some fresh approach, introducing new concepts and abruptly changing the course of the development. Ehrenfest's unique ability as a theoretical physicist lay neither in calculation nor creation; it lay in criticism, in "his unusually well developed faculty to grasp the essence of a theoretical notion, to strip a theory of its mathematical accoutrements until the simple basic idea emerged with clarity," in the words of Albert Einstein.[29]

Paul had put that talent for criticism to good use in tackling Boltzmann's papers. Boltzmann hadn't always been very good at explaining the mathematics he used or at mounting a cogent defence in the face of some quite harsh criticism from other physicists. At the time, few physicists were also mathematicians, which made it more difficult for them to follow the work of a fellow scientist who relied on unfamiliar mathematics. Most European physicists, including Mach,[30] were experimentalists who did not feel competent to handle complicated mathematical arguments. Perhaps that was one reason Boltzmann had often faced such resistance to his work. Of course, there were gaps in some of Boltzmann's theories, and they were sometimes weakened by flawed arguments. It took Paul's mathematical skills, his talent for explication and his ability to work away at a complicated idea or concept until it made sense to him to put Boltzmann's ideas on a sounder footing.

The encyclopedia article, noted Ehrenfest's biographer, was intended to make statistical mechanics understandable. "The encyclopedia article is written in his characteristic style, each section making a point as sharply as possible. There is no padding, no excessive verbiage. There is also a minimum of mathematical machinery; only a few key equations are given, and no calculations or derivations. This style works well for exposing a conceptual structure."[31] The publication of the article was very timely, as it drew Lorentz's attention to Ehrenfest just when he needed a job.

As interest in the quantum was beginning to bloom in the physics community, physicists looked to Boltzmann's statistical mechanics to understand the basis of the new quantum science. Paul and Tatyana's book-length article filled the bill very nicely.

Ehrenfest, however, did not give himself much credit for his talents. Instead, he longed for the creativity and genius that continually eluded him, and the scientific successes of his friends and colleagues served only to magnify what he had not achieved. As much as Paul looked to Tatyana to debate ideas with him until he could work through them to his satisfaction, he also depended on her to coax and cajole him out of his all too frequent bouts of dark self-doubt.

For Paul Ehrenfest, Russia was not all about the encyclopedia article and his depression. Just before he and Tatyana had arrived in St. Petersburg in 1907, a Swiss fellow, Albert Einstein, had published an interesting article on quantized light particles that captured Paul's interest. The article was based on a pivotal paper Boltzmann had written in 1877, the same paper Max Planck used as the basis for his 1900 paper on quantized radiation—the very beginning of quantum physics.

Chapter Four

A PLACE TO BELONG

Ambition . . . the glorious fault of angels and gods.
ALEXANDER POPE

When Hendrik Lorentz decided to retire from his position as chair of physics at the University of Leiden in 1912, he offered the position to Paul Ehrenfest, a timely and much-needed boost to Ehrenfest's nascent career.

Ehrenfest's lively style may have gone over well in St. Petersburg, but it was a little unorthodox for the restrained and formal atmosphere in Holland. Biographer Martin J. Klein described Ehrenfest's inaugural lecture at the University of Leiden:

> His vivid images and highly unprofessional language, his lively gestures and mobile face, all combined with the particular clarity and structure of his exposition to make a single unforgettable experience. It was something very new and thrilling, especially for the reserved, formal, academic society of the Dutch university, accustomed to the classic polish of Lorentz's lectures.[1]

Ehrenfest was a passionate man with the ability to sweep others along with his enthusiasm. With his dark eyes, thick mop of black hair and bushy beard, he looked more like a short, intense Russian soapbox orator than the youngest son of a working-class Viennese shop owner—or the chair of physics at a rather conservative Dutch academic institution.

He wasted no time in shaking up the university. Within weeks of his arrival in Leiden, he'd already started a regular Wednesday evening physics session and resurrected the moribund Christiaan Huygens Society as a gathering point for student mathematicians, chemists, astronomers and physicists. And he'd begun lobbying for space and funds for a reading room where physics students could have access to the collected works of the masters of their science as well as the most recent journal publications. The Wednesday evening colloquia soon became a staple in the diet of the Dutch physics community, and with funds from a private donor and a little genteel arm-twisting of administrators by Lorentz, the University of Leiden soon had its very first reading room. It was hard to resist Ehrenfest when he got the bit between his teeth.

The position in Leiden was heaven-sent. It was Paul's first real job, but he was taking over the reins from a revered scientist. He had a great deal of difficulty coping with the nagging worry that he was not up to the job and that Lorentz had made a terrible error in selecting him for the position. Not long into the job, he wrote to one of his friends in St. Petersburg about his feelings of inferiority:

> I don't have the mad idea that I should be providing an imitation of Lorentz. I know very well: [Leiden] is a little university in Holland—to which, by chance, Lorentz came as a high school teacher, and then just stayed. His successor had to be a really first rate fellow; they couldn't get one to come here, and so they had to take on a second rater, and I know that I am really in the forefront of the second raters right now.[2]

In fact, Ehrenfest *had* been the second choice as Lorentz's replacement. Lorentz had wanted Einstein, but Einstein had just accepted a position that allowed him to leave Prague and return to Zurich. For most physicists, being second choice to Einstein would hardly be construed as a slap in the face, but Ehrenfest couldn't shake the nagging belief that Lorentz had made a mistake by choosing him.

If Ehrenfest was Lorentz's second choice to replace him in Leiden, Ehrenfest was Einstein's first choice to replace him in Prague in 1912 when he wanted to move to Zurich. Unfortunately, Ehrenfest was not eligible for a university position in Austria until he formally declared a religious affiliation, and he and Tatyana had already declared themselves to be without one. Ehrenfest refused to change his religious status, much to Einstein's annoyance, since Einstein considered it nothing more than a bureaucratic formality. It may have been the first time the two men disappointed each other, but it certainly wouldn't be the last. Still, they had formed a close bond right from the time they first met, when Ehrenfest had been touring European universities a year earlier looking for a job.

Before leaving St. Petersburg in January 1912, Ehrenfest had written to Einstein to arrange a meeting, and Einstein had promptly invited him to come and stay at his home. Einstein and his wife, Mileva, met Ehrenfest at the train station and they headed to a coffee house to get to know each other. It was not until Mileva left to go home that their talk turned to physics. According to biographer Klein, before long the two were having an animated discussion about problems with statistical mechanics, walking in the rain to the university and arguing. That evening, Einstein had to leave to play in a string quartet, but when he returned home the two sat up arguing until the wee hours. They quickly formed a mutual admiration society, enjoying their passionate discussions and musical duets, where Einstein played violin and Ehrenfest piano.[3]

Ehrenfest didn't know then that the Einstein marriage was disintegrating, and that what had started out as a romantic and scientific partnership, much like the one he shared with Tatyana, had hardened into discord bordering on hostility. Part of the problem was Mileva's moodiness, but much had to do with Einstein's roving

eye and his restless quest to find a welcoming home and the nour-
ishment of undemanding adoration.

Einstein had grown up in the heart of a multiple-family household in Munich. His father, Hermann, and his father's brother
Jakob ran a business together, and their respective families all lived
in the same big house. Einstein's mother, Pauline, was the matriarch
of the family. She had grown up in a well-to-do multiple-family
household herself, and she knew how to negotiate the tricky shoals
of family dynamics and prevent unnecessary discord. Pauline was
very strict with young Albert, wanting him to be tough and independent. She had big ambitions for her little boy and she was
determined he was going to make something of himself—unlike
the amiable Hermann, who didn't have much of a head for business.

When Albert was fifteen, his extended family packed up and
moved to Italy to begin yet another business venture, leaving him
behind in lodgings to finish high school. For all intents and purposes, his big, noisy family had abandoned him. Then, to add
insult to injury, Albert's family home was sold to a developer, who
promptly tore down the trees and put up a row of apartment buildings where he and younger sister Maja and their cousins had
played. It wasn't long before Albert, depressed and lonely, quit
school.4

He wanted to study physics at the Swiss Federal Polytechnical
School (ETH) in Zurich, but first he had to finish high school. For
this, he boarded with the Winteler family in a town about twelve
miles from Zurich. "Papa" and "Mummy" Winteler, as Albert
called them, more or less adopted a quite willing young man into
their family of four sons and three daughters. Albert once again
found himself cosseted by a welcoming family, particularly by the
warm-hearted Mrs. Winteler and by nineteen-year-old Marie.

The Winteler and Einstein families expected Albert, then seventeen, to marry Marie, and so did she. However, once Albert
started university in Zurich in 1895, the romance cooled. Albert
had found a new romantic interest in Mileva Marić, a Serbian

woman who was in his physics classes. It is telling that when Albert
finally ended his relationship with Marie, he wrote not to Marie but
to her mother. It was as if he was more distressed about breaking up
with "Mummy Number Two," as he affectionately called her, than
with Marie. He excused his behaviour as the result of his devotion
to the "relentlessly strict angels" of science.[5] But Albert could not
really break up with the Winteler family or escape the jilted Marie.
His sister Maja had also moved in with the Wintelers to finish high
school, and his cousin Albert Koch boarded there as well.[6]

Like Tatyana Ehrenfest, Mileva Marić had had to leave her
home country in order to continue her studies after high school,
and she chose Switzerland. She was the only female enrolled in the
program to train mathematics and physics teachers at the ETH in
Zurich, the same program Einstein was taking.

Even in his teens, Albert could be unkind in his romantic relationships, especially if he felt he was being crowded or controlled.
Several years after they started seeing each other, Albert promised
Mileva he would not be spending as much time with the Wintelers,
since Marie had returned home from her teaching job and he didn't
want to risk running into the woman with whom he had once been
so madly in love. "For the most part," he told Mileva, "I feel quite
secure in my fortress of calm. But I know that if I saw her a few
more times, I would certainly go mad. Of that I am certain, and I
fear it like fire."[7] If Mileva had been feeling at all insecure in their
relationship, such "honesty" would not have been reassuring.

It was, perhaps, Einstein's deep attachment to his own family
that led to his more or less abandoning Mileva at a very difficult
time in her life, as well as his desire to avoid dealing with the emotional difficulties of his romantic alliances.

Albert and Mileva did not live together during their years at the
ETH for the sake of appearances, but they did have an intimate
relationship. Albert received a modest monthly allowance from his
mother's sister Julie to pay for his rooms. He did not live a lavish
lifestyle. He didn't drink, although he did indulge his penchant for

tobacco, often at the expense of his diet.[8] The two students rather fancied themselves as partners and scientists, just the two of them against the world, but it was more of a romantic illusion than a reality.

When exam time came in 1900, Einstein passed and Mileva failed. She had one opportunity to retake the examinations the following year, while also doing her doctoral dissertation—and then she found out she was pregnant. In the summer of 1901, with exams approaching, it was do-or-die time: she had to pass the exams or there was no point in continuing with her dissertation. She was unmarried, pregnant—and alone, as Einstein had chosen to go off on a summer holiday with his mother and sister. He preferred to be with his family rather than Mileva. She understandably failed the exams again, and that was the end of her ambition to become a mathematics teacher.

Mileva left Zurich and went home to her parents to have the baby. Einstein did not visit his very pregnant girlfriend. Rather, he spent his Christmas holidays with his mother and sister. About a month later, Mileva gave birth to a baby girl. Again, he did not visit, and it was understood that when Mileva came back to Zurich she would not be bringing the baby.* In fact, there is no indication that Einstein ever saw his daughter.

For his part, Einstein had found his four years studying at the ETH disappointing. His professor, Heinrich Friedrich Weber, had a traditional German background in theoretical physics and had

---

*What actually happened to Mileva and Albert's daughter Lieserl remains a mystery. She may have been given up for adoption or she may have died of scarlet fever. However, when a woman approached Einstein in 1935 claiming to be his daughter, Einstein hired a private investigator. The woman was eventually shown to be a fraud, but Einstein's reaction suggests either that he knew Lieserl had not died as a baby or that the woman was a fraud attempting to cash in on his fame. Or perhaps Einstein suspected that one of his many liaisons could have produced an out-of-wedlock child. (See Highfield, Roger, and Paul Carter, 1993, p. 93.)

been Helmholtz's assistant in the 1870s. To Einstein's dismay, Weber "simply ignored everything that came after Helmholtz,"[9] so, rather than attend classes, Einstein studied Maxwell, Lorentz and Boltzmann on his own. His fellow student Marcel Grossmann kept meticulous lecture notes, which he lent to Einstein so he could take and pass his exams. None of this endeared Einstein to Weber. At the end of his time at the ETH, Einstein was the only one of the four graduates not immediately offered a position at the ETH as an assistant, which Einstein put down to Weber's doing. The whole experience soured him on physics, but after about a year he regained his earlier interest and set to work writing a doctoral thesis.[10] Since the ETH didn't offer doctoral degrees, he submitted his thesis to the University of Zurich in the fall of 1901, where it was promptly rejected.[11]

After graduation, he had been unable to get a job at a university and he was growing desperate. Marcel Grossmann came to the rescue again. With the help of Grossmann's father, Einstein finally landed a position as a patent examiner in Bern in June 1902. Einstein had renounced his German citizenship when he was a teenager, in part to avoid the country's mandatory military service, and had only recently acquired Swiss citizenship. One of the reasons Einstein may not have been particularly keen to have it known that he had fathered an illegitimate child was that it could have put his new job at risk. The job was a civil service position in the Federal Department of Justice and Police,[12] so, for a while at least, Einstein appeared prepared to bow to convention and propriety.

A few months later, Hermann Einstein died, leaving Pauline Einstein saddled with a heavy debt load from her husband's failed business ventures, most of it owed to Pauline's well-to-do sister Fanny and her husband Rudolph Einstein, Hermann's first cousin. Pauline moved in with Fanny and Rudolph, but Albert still felt he had to do what little he could to help support her. In early 1903, Albert and Mileva married, but Mileva never quite seemed to recover from her failure to complete her training and the loss of her first-born child.

At this point in his life, there was very little to signal that Einstein was destined for a successful career in physics. Certainly his position as technical expert, third class, at the patent office did not portend great things. But Einstein may not have been too dismayed by the turn of events. He later said that his time at the patent office allowed him to think without the burden of the onerous career demands placed on young professors. "In a way," said Einstein, "this saved my life; not that I would have died without it, but I would have been intellectually stunted."[13]

Einstein had made several more attempts to have his revised dissertation accepted, without success. He had pretty much given up pushing for a doctoral degree once he took the patent clerk job, telling one of his friends that "it will be of little help to me, and the whole comedy has become boring."[14] Still, Einstein didn't quit. Instead, he made good use of his time at the patent office to ponder some of the puzzles and problems of the day in physics—not that he had a lot of spare time, given that he worked a 48-hour week. Einstein had already published five papers in the leading German physics journal *Annalen der Physik*, and he had obviously impressed at least one of the editors because he was invited to write reviews for the journal at the beginning of 1905.[15] But that's not all he was doing.

That same year, between March and September, Einstein wrote five more papers, all of which were published in *Annalen*. In the summer he took another stab at a doctoral degree, submitting his June paper with its radical new theory of relativity. It was promptly rejected by the Zurich faculty. He then tried his less controversial April paper on deriving the size of sugar molecules. That too was returned because it was considered too short to warrant a doctoral degree. Einstein is said to have then added a single sentence to the paper and resubmitted it. It was accepted. Albert Einstein finally had a doctoral degree.[16]

Einstein had been cultivating a "lone wolf" persona ever since he was a teenager, although at times it appeared merely a convenient excuse for distancing himself from the importunate females and domestic demands in his life. According to biographer John Rigden, Einstein's preference for working alone allowed him a certain independence in his thinking:

> Isolated from active working physicists as he certainly was, Einstein became accustomed to working alone. For the most part, he continued to do so throughout his career. He was the sole author of the 1905 papers and most other papers. Because the approach he took to physics was singular, neither credit nor blame could be shared. With no co-author dabbling in this work and with confidence in his own insights, he had no doubts about the correctness of his work.[17]

Even though he was out of the loop academically, Einstein was proving that he could come up with original ideas in physics. In 1904 he had come across *Science et l'Hypothèse*, a book by noted French mathematician and physicist Henri Poincaré published in 1902, in which Poincaré identified three unresolved physics problems: the cause of the zigzagging motion of particles in liquid, called Brownian motion; the lack of experimental evidence for the ether; and an explanation of the photoelectric effect. In 1905, Einstein resolved the question of Brownian motion in his May paper, did away with the necessity of ether in his June paper on special relativity, and suggested that the photoelectric effect might experimentally offer verification of the light quanta hypothesis he presented in his March paper, called "On a Heuristic Point of View about the Creation and Conversion of Light."

Scientists at the turn of the century had noticed during lab experiments that light shining on a metal appeared to be knocking electrons out of the surface of the metal, but the kinetic energy of these "photoelectrons" seemed to be somehow fixed. A brighter light would knock out more electrons, but the freed electrons didn't move any faster. The accepted wave theory of light offered no explanation for this behaviour.

Einstein hadn't actually been attempting to solve the photo-electric problem; that was just a side effect. Rather, he was using the statistical interpretation of the second law of thermodynamics to support his argument for light quanta, and for this he naturally turned to Boltzmann's 1877 paper for inspiration. He was familiar with Boltzmann's statistical mechanics, having already published three papers on the subject during the period from 1902 to 1904.[18] Visualizing Boltzmann's imaginary container of gas molecules, Einstein posed the question of how the entropy—the measure of disorder of the gas molecules—would change if the volume of the container got smaller but the number of molecules stayed the same, and he proposed that the change in entropy corresponded to the change in volume.

Einstein avoided making any assumptions about entropy or the structure of gas molecules. Rather, he limited his assumptions to two simple requirements: that the gas particles in the container moved independently or randomly, and that they retained this random motion regardless of the change in volume of the container. From what was known about the physics of an ideal gas—how gas behaved under ideal but imaginary conditions, as opposed to messy real conditions—these were quite reasonable assumptions.

Then Einstein asked another question. What would happen if the size of an imaginary container of radiation—the kind that Planck and other physicists had been employing to figure out the spectral distribution of radiation—changed but the energy of the radiation inside stayed the same? Again, from the known behaviour of the entropy of radiation, he argued that the resultant change in entropy corresponded with the change in volume.

Having shown that the equation for the change in entropy for a container of gas was the same as that for a container of radiation, he pointed out that radiation behaved "as if" it were composed of independent energy quanta.[19] If radiation could be assumed to be "light particles" or quanta, it could be treated as if it were a collection of gas molecules. Thus, Einstein was suggesting that light (radiation) behaved as though it had the characteristics of a collection of particles.

If that were all Einstein had demonstrated in his March 1905 paper, he still would have made a quite significant contribution to physics by at last developing a theory that embraced radiation, particles and heat, showing how the three were related to each other. But he went one step further, citing three possible experimental supports for his new light quanta hypothesis, one of which was the photoelectric effect.[20] This problem, said Einstein, could be explained if the light hitting a metal surface were made up of light particles or quanta rather than light waves.

Einstein wasn't the first person to notice that light had properties of both particles and waves. Newton believed light was a stream of particles, but experiments that showed two light waves interfering with each other like two circles of waves in a pond had cemented the validity of the wave theory. According to biographer and historian of science Martin J. Klein:

> In this paper, Einstein set himself against the strong tide of nineteenth century physics and dared to challenge the highly successful wave theory of light, which was one of its most characteristic features. He argued instead that light can, and for many purposes must, be considered a collection of independent particles (quanta) of energy that behave like particles of gas.[21]

Einstein's paper, however, had little impact on mainstream physics. His arguments were both simple and daring, rooted in thermodynamics and statistical mechanics, but that didn't mean he could convince his colleagues of the validity of his light quanta. As Klein noted, "Very few were willing or able to follow him accepting the startling idea of light quanta on the strength of deductions that were based on the statistical interpretation of the second law of thermodynamics."[22] Perhaps the strongest evidence that few physicists had bothered to read Einstein's 1905 paper was the assumption that he had based his quantum hypothesis on Planck's introduction of quanta.[23] Einstein was certainly familiar with Planck's quantization, but he used a quite different approach.

By the end of 1905, Einstein had his doctoral degree, had produced a handful of papers that addressed key issues of the day, and had written twenty-one reviews and abstracts for *Annalen*. If this busy year offered Einstein significant professional satisfaction, the year following brought profound personal tragedy. In the winter of 1906, one of the Winteler boys, recently returned from America, went mad and shot "Mummy" Winteler, the husband of one of the Winteler girls, and then himself.[24] The double murder and suicide brought to a sudden and harsh end another of Einstein's idealized happy families.

Even if few other physicists were bothering with Einstein's radical light quanta theory, Ehrenfest was paying attention. Like Planck's energy quanta paper of 1900, Einstein's 1905 paper (as well as his earlier ones) was based in Boltzmann's statistical mechanics, and that was Ehrenfest's specialty.

Both Planck and Einstein continued to publish papers on their respective quantum theories, and once Ehrenfest was finished with the encyclopedia article on statistical mechanics in 1910, he turned his attention to the problem of the quanta. This was something he'd been studying for nearly a decade, puzzling over the meaning of Planck's strange idea of quantization and the even stranger ideas of Einstein.[25] Ehrenfest thought it would be a good idea to report on some of the work he'd done tying together Einstein's light quanta and Planck's energy quanta so that "at least some of the features of the quantum hypothesis could be viewed as established once and for all." He did this in an article entitled "Which features of the hypothesis of light quanta play an essential role in the theory of thermal radiation?"[26]

This was the first comprehensive review of the early, awkward stages of quantum physics, and it attracted almost no interest. Ehrenfest's paper was published in the journal *Annalen der Physik* in October 1911, shortly before the First Solvay Conference in Brussels. The theme of the conference was "Radiation Theory and the Quanta." Ehrenfest's paper would have fit nicely into the proceedings, particularly since it clarified some of the issues to be debated at the conference. But Ehrenfest was an unemployed physicist living in St. Petersburg at the time, and he hardly warranted an invitation to a gathering of such notables as Hendrik Lorentz, Max Planck, Marie Curie, Paul Langevin, Ernest Rutherford, Henri Poincaré and Albert Einstein, the youngest of the invited scientists.

Poincaré, a man of major stature in the European science community, did not have a background in the quantum, but he was so inspired by the discussion at the conference that he went back to Paris and wrote his own analysis of the quantum. His article, dealing with many of the same issues that Ehrenfest had written about, was published in the *Journal de Physique* in January 1912.[27]

Poincaré had not seen Ehrenfest's paper in the German journal, and Ehrenfest did not see Poincaré's paper in the French journal until he was visiting universities in Germany and Austria in the spring of 1912 in search of a job. Ehrenfest sent Poincaré a copy of his paper, which Poincaré promptly acknowledged and said he would use in his next article. Unfortunately, Poincaré died a few months later without publishing the promised paper that would have drawn attention to Ehrenfest's work. Thus, it was Poincaré and not Ehrenfest who was credited with awakening the larger science community to the intriguing possibilities of quantum physics.[28]

About the same time, 27-year-old Niels Bohr was organizing his move from Cambridge to Manchester to work on radioactivity with Ernest Rutherford. Bohr had considered working with Lorentz in Leiden after he finished his doctoral thesis in 1911, but had opted instead to work with J.J. Thomson in Cambridge. But Thomson had little patience for or interest in the young man who could barely speak English. Bohr had been impressed by a lecture Rutherford had given at Cambridge, so he made a point of visiting him in Manchester. Rutherford had just returned from the Solvay conference and was excited about having met Einstein and Planck. He regaled Bohr with stories of what had happened in Brussels. Bohr was so inspired by Rutherford that he decided to move to Manchester, keen to start working on Rutherford's radical new model of the atom, the one with electrons whirling around a nucleus like a miniature solar system.

Chapter Five

BUILDING A FOUNDATION

Sometimes I think we don't solve anything.
We just rearrange the mystery.

TONY HILLERMAN, *Coyote Waits*

The Ehrenfest living room was often filled with noisy debate about the latest physics discovery. The comfortable sofa and chairs routinely seated the physicists and students who attended Ehrenfest's invitation-only colloquia or the many visiting scientists who enjoyed the Ehrenfests' hospitality.

Tatyana herself had designed their big sunny house at the end of Witterozenstraat, with its views of the garden and the adjacent canal.[1] The library/study/workroom stretched the whole width of the house, with three windows looking out onto the garden. A large blackboard was propped up on a sturdy easel at one end of the room, a most useful device for the colloquia and for enthusiastic scribbling of equations that set the chalk dust flying. On the walls hung pictures of St. Petersburg, reminders of the Ehrenfests' years in Russia, and books by Tolstoy and Dostoyevsky could be found amongst the science texts in the bookcases. The library/study was also the family living room and the place where guests were entertained. In its own way, the room represented a family lifestyle where scientific discourse and daily living overlapped. It was a room intended for discussion and debate, not for reflective solitude.

Jan Burgers, one of Ehrenfest's first students, recalled:

To me, this room has been more impressive than the study rooms of any other scientist by whom I have been received. It had a simplicity and austerity, and for me it has a grandeur, since it speaks of the many chapters of modern physics which have been discussed in it, with Ehrenfest's colleagues [including his wife!] as well as many, many visitors who have come to see him. The room has spoken to me the stronger since I was received here as a close friend, almost as a close relative, while also my best friends were at home there.[2]

It was not surprising to find Tatyana Ehrenfest at colloquium gatherings, even allowing that they were held in her own living room. She was not a physicist, but she was an excellent mathematician, and with a mathematician's logic she could—and would—point out flaws in reasoning or gaps in logic in others' arguments, just as she did for Paul.

The Ehrenfest household was a busy one. Three of the children were home-schooled by a governess. Paul had been quite adamant that his children would be spared his miserable experience in the formal school system, a difficult time in his teens that appeared to have exacerbated his doubts about himself and his abilities. Their youngest son, Vassily, had been born with Down's syndrome and was severely disabled. Paul's brother Hugo, the doctor, had recommended the child be institutionalized, and Vassily was placed in a facility in Jena.[3]

Paul and Tatyana had sunk most of their money into the house, and with the fees for Vassily's care, money was tight. Most European professors earned modest salaries, and Paul was no exception. To supplement his salary he conducted state exams, an institution he detested, and he charged fees for giving public lectures. But his main source of extra income came from the Philips Laboratory, a company that employed a large number of Dutch physicists.

Ehrenfest's job was to organize a series of lectures by some of Europe's top physics theorists and experimentalists to help keep the Philips lab people up to date.[4]

Visiting guests often stayed with the Ehrenfests; it was common practice for physicists to stay with colleagues who were also friends when they were in town to collaborate on research, give guest lectures or attend conferences. Often, a visiting scientist would arrive for a few weeks' stay with an entourage, and his hosts might be obliged to find accommodation for a spouse, a child or two, perhaps a governess or maid, and sometimes even pets. It could create a lot of household chaos, and some spouses found such visits to be an excellent time to take the air in the mountains or leave on a seaside vacation with the children.

While hospitable, Paul and Tatyana ran a somewhat unorthodox household, and not just because their living room doubled as a physics lecture hall. Their household was vegetarian, and no drinking or smoking were allowed in the house. If guests fancied *frikandellen* (fried sausages) and a pint or two of ale, they could take themselves off to one of Leiden's taverns. If they wanted to smoke, they could go out to the garden.

Einstein adored the Ehrenfest family. He was more than happy to plant his feet under their kitchen table, even if there was no meat dish being served. Einstein and Ehrenfest had become close friends, and Ehrenfest, who wanted to see more of Einstein, was trying to coax him to move from Berlin to Leiden. In 1919, on behalf of his university colleagues, Ehrenfest made Einstein an attractive offer: the maximum pay, no necessity to teach and no need to learn Dutch to give lectures. He was insistent that people at the university wanted Einstein there because they were fond of him personally, "and not just of the brain drippings that ooze out of you!"[5]

Einstein turned down the offer, mainly because he felt obligated to stay in Berlin unless conditions there became untenable. He'd made that promise to Planck when he moved there. Undeterred, Ehrenfest then offered him the position of special visiting professor in Leiden, which offered the opportunity of staying with the Ehrenfest family for several weeks each year. Einstein accepted.

With the difficulty of travelling during the later war years, it was three years since the Ehrenfests had last seen Einstein when he finally arrived in October 1919. Einstein made it to Leiden just after Lorentz telegraphed him that his theory of general relativity had been verified by the Eddington expedition. This news gave Paul and Albert lots to talk about: Ehrenfest's plans for Einstein's impending visiting professorship and the validation of relativity.

On his return to Berlin after his two-week visit, Einstein wrote a heartfelt letter of appreciation to Paul, Tatyana and the children for "the beautiful and tranquil time that we spent together" and for the attention to his health and comfort. "I have never before taken part in such a happy family life," Einstein wrote. "It comes from two independent people who are not united just by compromises. I have come to feel that all of you are a part of me and that I belong to you." The reply from the normally ebullient Ehrenfest was brief but equally heartfelt. "For your answer you must simply imagine the children's faces, full of love for you."[6]

Einstein had found himself another big, welcoming family. Unfortunately, as much as he wanted to be part of such a family, he seemed to have little ability to create one of his own, despite years of trying. But perhaps he hadn't tried all that hard.

While he was still married to Mileva, Einstein had taken up with his first cousin Elsa, the eldest daughter of his aunt Fanny, his mother's sister. And since Fanny had married his father's first cousin, Albert and Elsa were also second cousins, which made Elsa's daughters from her first marriage, Ilse and Margot, Albert's second and third cousins. When Mileva had left Albert in 1914 and moved back to Switzerland with their two boys, Albert had been happy enough to live on his own whilst enjoying the care and ministrations of Elsa, who anticipated becoming the next Mrs. Einstein. Then, when Albert collapsed in 1917 from chronic stomach problems and overwork, he'd needed someone to care for him. He moved into the apartment next to where Elsa lived with Ilse and Margot, and then finally moved in with them. The arrangement had the potential to become a happy family situation, especially given that they were already family—but it didn't.

The romantic relationship between Albert and Elsa had cooled by 1919 and soon became a union of convenience, but Elsa still expected Albert to marry her. She had, after all, put many years into cooking, cleaning and caring for him, and if he didn't marry her, she faced the humiliation of being demoted to the status of his housekeeper. Albert and Elsa did finally marry on June 2, 1919, in a registrar's office in Berlin, with the understanding that Elsa had no claim to Albert's fidelity. Perhaps it was the contrast between the chill of his own household in Berlin and the warmth of the Ehrenfest home in Leiden that led to Einstein's outpouring of affection for the Ehrenfest family when he visited them a few months after his marriage.

Ehrenfest had the happy family, but he lacked the creative genius of his two dearest friends. Einstein had the creative genius, but little ability to sustain a happy family of his own. Niels Bohr, on the other hand, seemed to have it all.

Bohr was born into a Danish family considered "high to the ceiling," meaning a family of rarefied intellectual standing and goodness.7 Niels was also part of an academic dynasty stretching back to his great-grandfather. His father and grandfather were both professors, and there was never any question that Niels would be one too. His mother was the daughter of a Jewish banker and parliamentarian, and his father was considered to be "one of the men around whom the intellectual and cultural life of Copenhagen revolved."8

Niels was born in the family mansion facing the Christiansborg Castle, and when his father became a university lecturer in physiology in 1886, baby Niels and his older sister Jenny moved with their parents into an apartment in the gracious old Academy of Surgery building. Brother Harald was born shortly thereafter, and the children grew up in the stately building. Once the boys were old enough, they were allowed to sit in on the discussions when some of the greatest scientists and leading philosophers of the day came to visit their father.

Niels was only a year and a half older than Harald, and the two brothers were very close. As they grew up, the handsome brothers became celebrity soccer stars. Harald was the better player, and he made the Danish Olympic team in 1908, the year Denmark won the Olympic silver medal for soccer. But both Bohr boys were considered national celebrities, with newspapers reporting on their exploits on and off the soccer field.[9]

At university, Harald was studying mathematics, Niels physics. The problem for Niels was that, when he started working on his master's degree in 1903, there was only a single theoretical physicist at Copenhagen University, and almost nothing by way of laboratories or laboratory equipment for physics experiments. Denmark was definitely lagging behind other European countries in supporting physics research.[10] During the ten years from 1901 to 1910, only a dozen master's degrees were awarded in physics or mathematics from Denmark's lone university,[11] and two of them went to the Bohr brothers. Harald and Niels may have been part of an academic family dynasty, but there was definitely no physics or mathematics dynasty in Denmark. With few advanced physics courses available to him, Niels was left a lot of the time to work on his own, in much the same way Einstein had worked independently at the ETH.

In 1910, Harald finished university. It was customary in Denmark for a doctoral defence to be a white-tie-and-tails affair, open to the public. The Danish newspapers, which covered the event, reported that the well-known soccer player's mathematics thesis defence drew a most unusual crowd comprising mainly soccer players.[12]

Before starting work on his own doctoral thesis, Niels took a short summer holiday that same year, but it was long enough for him to fall in love with the pretty blonde Margrethe, the sister of one of Harald's friends. Harald went off to Göttingen University to study with the famous mathematicians there. To give Niels the seclusion he needed to prepare his thesis, Bohr's father arranged for him to stay on the tranquil island of Funen, the setting for Hans Christian Andersen's fairy tales.

According to biographer Abraham Pais, Bohr had already established a pattern in the preparation of his master's and doctoral theses of laboriously writing many drafts of his papers, sometimes hundreds of pages, before handing them to someone else to write— or dictating them. Sometimes it was his mother who transcribed his papers, sometimes Harald. By the time of Bohr's doctoral thesis, Niels and Margrethe were engaged and she was contributing her fine penmanship to his papers.[13]

Niels's public thesis defence on "Studies in the Electron Theory of Metals" in May 1911 drew just as much public attention as Harald's had. The next day, one Danish newspaper reported:

> Here in Denmark there was hardly anybody well informed enough about the electron theory of metals to be able to judge a thesis on this subject.
>
> Dr. Bohr, a pale and modest young man, did not take much part in the proceedings, the short duration of which is a record. The little Auditorium III was overflowing and people were standing out in the corridor of the university.
>
> The words Bohr had written and the questions he raised were literally so new and unusual that no one was equipped to question them.[14]

Not long after the thesis defence, Niels left his fiancée behind to work in England with J.J. Thomson, the Nobel laureate credited with discovering the electron. Since there were so few physicists in Denmark, Bohr was accustomed to working mostly alone, reading Poincaré, Thomson, Lorentz, Planck and Einstein, so the brief time he spent in Cambridge was his first real experience of working within a larger physics community. Unfortunately, Thomson's lack of interest in the young Dane made his time there of little use to him; his subsequent sojourn in Manchester, working under Ernest Rutherford, was much more fruitful.

Rutherford, who grew up in the rugged backblocks of New Zealand farm country, had been working on radioactivity at McGill University in Montreal before leaving Canada for Manchester in 1907. The following year he was awarded the Nobel Prize for his work in chemistry. By 1911, Rutherford had taken Thomson's "plum pudding" model of the atom—where negatively charged electrons inhabited some sort of positively charged blob, like plums in a pudding—and modified it so that the electrons orbited a tiny, positively charged nucleus.

Manchester was the leading centre for experimentalists studying radioactivity, and their experiments had produced supporting evidence for Rutherford's nucleus with orbiting electrons. Bohr wasn't particularly interested in radioactivity, however; it was electron theory that had drawn him to England. Curiously, Rutherford's orbital model of the atom attracted little attention in Manchester. Although he had come up with the model before heading off to Brussels for the First Solvay Conference in 1911, Rutherford stayed quiet about it, as if he wasn't yet really sure about his new model of the atom.

Funding for Bohr's position in Manchester lasted only until July, giving him just three months to develop some understanding of this new atomic model and some insight into the problems the experimentalists he met there were having in understanding their results. Bohr was working on a paper about atomic radiation when he hit upon the idea that the failure to match the existing theory with experimental results might be due to the assumption that electrons were orbiting the nucleus freely. He wondered if there were limits on how far electrons could be from the nucleus. Then, in

June, he suspected he'd hit upon a good idea. In much the same way as when Planck employed imaginary resonators with energy that was limited to the frequency multiplied by the constant h in his theory of quantized radiation, Bohr employed imaginary atomic vibrators for electrons, using the same kind of quantized energy.[15]

Developing this idea required the use of mathematics, which was not Bohr's strong suit. Harald was the mathematician in the family, and he had usually been on hand to help Niels when math problems came up. But now Harald was in Göttingen. "I have thought of you often these days," Niels wrote to his brother, "for I had to use some mathematics, and thought of asking for your advice."[16] In another letter to Harald that followed soon after, he wrote:

> It could be that I've perhaps found out a little bit about the structure of atoms. You must not tell anyone anything about it . . . You understand that I still could be wrong, for it's not completely worked out . . . You can imagine how anxious I am to finish quickly and have stopped going to the laboratory for a couple of days to do so (that's also a secret).[17]

With his time in Manchester almost over, Bohr wrote up a report of his ideas on the quantized orbits for Rutherford— although he didn't refer directly to quantization—so they could talk them over before Bohr left. In his memorandum, as he called it, Bohr explained the need to address a key problem with the Rutherford model of the atom: there was no mechanical basis for the stability of the atom, and no way to explain why the electrons didn't simply spiral into the nucleus as it radiated energy away.

> This hypothesis [he wrote], for which there will be given no attempt of a mechanical foundation (as it seems hopeless), is

chosen as the only one which seems to offer a possibility of explanation of the whole group of experimental results which gather about and seem to confirm conceptions of the mechanism of the radiation as the ones proposed by Planck and Einstein.[18]

Rutherford's reaction to Bohr's model of the atom was cautious to say the least. He warned his young colleague that there was little experimental support for his orbital model, and that Bohr shouldn't rely too much on it. In any event, further pondering of the quantized atom would have to wait, as Bohr headed back to Denmark in the summer of 1912 to get married, his paper on atomic radiation only half done. The wedding was a two-minute ceremony with the Copenhagen chief of police officiating and Harald, of course, acting as best man. Margrethe's mother did, however, insist on a large wedding banquet afterwards.

For their honeymoon, Niels and Margrethe went to Cambridge so that Niels could finish the radiation paper he'd started in Manchester. He dictated the paper to Margrethe, who wrote it down neatly, and then they went off to Manchester to deliver it to Rutherford in person.

Bohr may not have spent a great deal of time with Rutherford, but the hearty New Zealander had made a big impression on him. Not only did Bohr get to see first-hand how a well-run lab functioned, he also got a good look at how a very busy man handled the heavy load of running a facility while still coming up with innovative research ideas. Rutherford was a powerful personality who inspired his young collaborators to do their best possible work. "Although there was no doubt as to who was the boss," said one such Rutherford collaborator, "everybody said what he liked without constraint . . . He was always full of fire and infectious enthusiasm when describing work into which he had put his heart and always generous in his acknowledgement of the work of others."[19]

Bohr's opportunity for a physics job in Denmark had arisen just after he'd left for Cambridge the previous fall. The theoretical physicist who had been his teacher was set to retire, and his assistant was the obvious choice to replace him. That left the assistant's position open for Bohr. There were so few physics positions at Denmark's one university that if Bohr didn't get the assistant's job, he might have to wait a very long time for another opening. Fortunately, he was offered the job, which he accepted the day before his wedding on August 1.

Bohr wanted to work on his ideas about the atom, something that was of no interest to his new boss. So he asked to be relieved of his position in the fall of 1912, and he and Margrethe retired to the country to work on what was essentially an expansion of the Rutherford memorandum.[20] In particular, Bohr wanted to follow up on the work of a British astronomer whose paper, published that summer, on the spectrum of the solar corona utilized a model of quantized orbits for the rings around Saturn that was remarkably similar to his model of quantized electron orbital rings around the atomic nucleus.[21] The astronomer's results were nowhere near the predictions of Bohr's model—which is perhaps not surprising given the enormous difference in scale between atoms and planets—but the paper did open Bohr's eyes to the usefulness of spectral analysis.

Researchers had long known that light shone through a prism split into the colours of the spectrum, from violet to red. They also knew that when light radiated by heating specific elements, such as hydrogen and sodium, was shone through a prism, it produced a signature "fingerprint" that was unique to each element. The fingerprint was made up of certain bars of colour from the spectral rainbow, some bright and some dim (rather like a multicoloured bar code). In 1814, a German glassmaker and physicist, Josef Fraunhofer, examined sunlight directed through a slit into a prism he'd made and was able to see dark lines in the coloured spectra.

Nobody knew what those dark lines meant until nearly fifty years later.

In the 1850s, physicist Gustav Kirchhoff and chemist Robert Bunsen invented the spectroscope out of "a prism, a cigar box, and two ends of otherwise unusable old telescopes."[22] Suddenly it was possible to get a much clearer picture of the spectral signatures and to match elements with their respective coloured-line fingerprints. Kirchhoff and Bunsen were able to show that the dark Fraunhofer lines were simply gaps in the prismatic array. By matching the black lines in the spectrum of sunlight with the coloured lines of the hydrogen spectra, for instance, scientists could tell that the sunlight had passed through a hydrogen cloud which had absorbed those specific colours. Nobody had any real idea why chemical elements had unique spectral signatures, but it turned out that those signatures could be used to identify the chemical composition of everything from mineral water to stars to solar coronas.

Then, in 1885, a Swiss schoolteacher by the name of Johann Balmer came up with a formula that correctly predicted the lines of colour in the hydrogen spectrum, but still the origin or cause of the spectral fingerprints remained shrouded in mystery.

Bohr didn't learn of the Balmer formula for the series of spectral lines for hydrogen until early in 1913,* but he quickly realized that he could make use of it to check his quantized orbits for the electron in the hydrogen atom and see if his model was workable. Within two weeks of discovering the Balmer formula, Bohr had come up with the long-awaited explanation for the hydrogen spectrum. He did away with the imaginary atomic vibrators that were supposed to absorb and emit radiation. Instead, he assumed that electrons, to which he had already assigned quantized orbits, absorbed and emitted radiation of a certain frequency as they passed from one allowed orbit to another. The frequency of the radiation

---

*Since the Balmer series was noted in a textbook written by the Danish physicist who had been Bohr's teacher, it is possible Bohr had simply forgotten about it.

for the transition of an electron between two of Bohr's orbits for a hydrogen atom matched exactly the frequency of the coloured lines of the hydrogen spectrum given by the Balmer formula. Indeed, one had only to subtract the energy of one allowed hydrogen orbit from another orbit and divide the result by Planck's constant *h* to arrive at the frequencies in the Balmer series.

Bohr sent the draft of his paper off to Rutherford in early March, and he replied promptly with a point that worried him. "There appears to me one grave difficulty in your hypothesis," Rutherford wrote to Bohr, "which I have no doubt you realize, namely how does an electron decide what frequency it is going to vibrate at when it passes from one stationary state to another? It seems to me that you have to assume that the electron knows beforehand where it is going to stop."[23]

Bohr didn't have an answer for that question. Besides, there was a fundamental problem with the solar system model for the hydrogen atom. Unless there is a balance between the gravitational and centrifugal forces in the solar system, planets will either spiral into the sun or go off into space. The atom had a similar problem. According to classical physics, an electron in a hydrogen atom should rapidly spiral into the nucleus. But that obviously didn't happen, because it would have made atoms too unstable to form the molecules of gases, bedposts and stars. Bohr simply imposed the condition that the lowest energy orbit—the one closest to the nucleus—was stable. The electron in that orbit could not spiral into the nucleus because its orbit did not allow it to get any closer. The electron had to either stay in that orbit or absorb energy to "jump" to a higher allowed orbit.

Bohr's paper on the hydrogen atom was published in the British *Philosophical Magazine* in July 1913. If Bohr had only proposed an atom with quantized orbits, the lowest of which he had decreed as stable, the idea might have been ignored or dismissed. After all, there was no experimental evidence to support the idea. But since there was lots of experimental evidence for the spectrum of hydrogen and other elements, Bohr's model opened the door to understanding the spectrum of the elements, and that was something that could not be ignored.

In September, Bohr published a sequel to the July hydrogen paper that suggested an explanation of the structure of other atoms and how the number of electrons in an atom matched its place in the periodic table. Another paper followed in November that dealt with the significance of the outermost ring of atoms in the structure of molecules. By December, Bohr had polished the results from the trilogy of papers and presented a final 1913 paper to the Danish Physical Society. It had been a very productive year.

The reaction to Bohr's model of the hydrogen atom in the latter half of 1913 was decidedly mixed. In England, J.J. Thomson simply ignored it; Rutherford worried about it. Harald wrote from Göttingen that there was a lot of interest in Bohr's work but also some doubt that such bold ideas could be right. Arnold Sommerfeld, the professor of theoretical physics at Munich, was impressed by Bohr's derivation of the Balmer formula but wasn't quite sure about his atomic model. A physicist in Berlin had, however, taken the unusual step of highlighting Bohr's July paper at a meeting of the Berlin Physical Society, calling it a "stroke of genius."[24]

Ehrenfest did not find Bohr's atomic model at all appealing, despite having a far better understanding of quantization than most other physicists. With his tendency for melodrama, he wrote to Lorentz, saying, "Bohr's work on the quantum theory of the Balmer formula (in the *Phil. Mag.*) has driven me to despair. If this is the way to reach the goal, I must give up doing physics."[25]

One of Bohr's experimentalist friends from his Manchester days reported to Bohr in the fall that Einstein thought his model was quite good. But Einstein wasn't following the development of the quanta very closely at that point. He'd written a number of papers advancing his theory of the light quanta up until 1911, but after that he'd become immersed in developing a generalization of his theory of special relativity.

Bohr had made use of the earlier work by Planck and Einstein to develop his model of the hydrogen atom, so his idea of the atomic model contained the same problem: there was still no physical meaning to Planck's *h*, nor was there any explanation for the "chunking" of energy in the microscopic world. Indeed, Planck was still trying to distance himself from the strange quantum, hoping it could eventually be reconciled with classical physics.

Ignoring Bohr's atom, Ehrenfest continued trying to figure out a way to establish some kind of basis for the boundary between the microscopic (quantum) world and the macroscopic (classical) world. By now, lots of physicists were making use of Planck's *h* to quantize problems in classical physics, sometimes just plugging *h* into an equation. The application of the quantum of action seemed quite arbitrary and it was difficult to see where it made sense and where it didn't, unless the application was obviously contradicted by experimental results.

Dealing only with the quantization contributions of Einstein and Planck, Ehrenfest came up with some limits for describing stationary quantum systems—his "adiabatic hypothesis"—and published a detailed paper on the subject in October 1916. The paper garnered little attention except from Einstein, who gave it serious consideration. Einstein had published his generalized theory of relativity the year before and had now turned his attention to quantum physics. Unbeknownst to Ehrenfest, Bohr had also considered the paper and applied it to the stationary states, as he called the allowed orbits, of the quantum atom.

Before the Great War, Bohr had accepted Rutherford's offer of a temporary position at Manchester, but by the summer of 1916, Niels and Margrethe were back in Copenhagen. The Danish government had approved the creation of Denmark's second theoretical physics position, and awarded the new job to Bohr. Not long after assuming his new position, Bohr agreed to take on a bright young student from Leiden who was studying for his doctoral degree in physics under the supervision of Paul Ehrenfest.

Hendrik Kramers became Bohr's first research student and relieved Margrethe of her writing duties, which was just as well because she had recently given birth to the Bohrs' first child.

Bohr incorporated ideas from Ehrenfest's 1916 paper into a lengthy paper of his own that he published in 1918, his first paper since 1915. There were a number of reasons for the dry spell. The only space Bohr had to work in was a tiny office, which he shared with Kramers, and he was busy planning and lobbying for a proper physics institute of his own. But Bohr was also frustrated by the way the new physics was evolving and changing before he could get his ideas down on paper. "My life from the scientific point of view," he said in 1918, "passes off in periods of overhappiness and despair, of feeling vigorous and overworked, of starting papers and not getting them published, because all the time I am gradually changing my views about this terrible riddle which quantum theory is."[26]

Ehrenfest didn't know about Bohr's paper until Kramers returned to Leiden for a visit, bringing with him a letter from Bohr to Ehrenfest. By then Ehrenfest's views on Bohr's atom had softened, and he was delighted when Bohr closed his long letter with the hope that he could visit Holland and meet Ehrenfest once the war was over.[27]

The Great War ended in the fall of 1918, and the following January, Ehrenfest invited Bohr to stay at his house while attending both an April conference of Dutch scientists and Kramers's doctoral thesis defence. Ehrenfest said Bohr was welcome to stay as long as he could stand the chaos of their household, and that he would endeavour to get Einstein to Leiden at the same time. Einstein and Bohr had not yet met, but it would have to wait. Einstein wrote to Ehrenfest at the end of March that he had delayed in responding to the invitation to Leiden because he was waiting to see if the street demonstrations and other post-armistice chaos in Berlin would settle down enough for him to get away. "I am mightily drawn to visit you," Einstein wrote. "On the other hand, travelling is dreadful, especially for someone with queasy intestines . . .

I would really like to get to know Bohr, with his marvellous intuitive gift. But it can't be done."[28]

Ehrenfest may well have been quite disappointed at not being able to introduce the two men, but he and Bohr had a good time anyway. Ehrenfest was an enthusiastic host, introducing Bohr to Dutch physicists, showing off the spring fields of tulip bulbs and touring museums, while at the same time making sure Bohr had the peace and quiet he needed in order to think. Like Einstein, Bohr appreciated the lively discussions in the Ehrenfests' study and having a passionate fellow scientist and friend with whom he could share personal interests. The likeable man Ehrenfest got to know during the days of Bohr's visit was not the man suggested by his laboured and difficult papers.

Bohr, in his own awkward way, was just as effusive in expressing his pleasure in Ehrenfest's company as Einstein had been. After his return to Copenhagen in May, Bohr wrote to Ehrenfest, "I miss so much to express my feeling of happiness over your friendship and of thankfulness for the confidence and sympathy you have shown me. I find myself utterly incapable of finding the words for it."[29]

When Einstein finally did make it to Leiden in the fall of that year, he heard a lot about the Bohr visit. Ehrenfest pitched Bohr's ideas to Einstein and sang the praises of his new-found friend. Intrigued, Einstein promised to bury himself in Bohr's papers on his return to Berlin. "You have shown me that there is a man of profound vision behind them," Einstein wrote to Ehrenfest, "in whom great connections come alive."[30]

Ehrenfest was not easily put off once he'd made up his mind, and he was quite determined to bring his two friends together, sooner or later.

Chapter Six

THE COST OF COMPROMISE

Everyone is entitled to his own opinion, but not his own facts.
DANIEL PATRICK MOYNIHAN,
American senator

Paul Ehrenfest had every reason to be pleased with himself in the fall of 1925. It was no small feat to get two of the world's most influential and sought-after physicists to sit down together in his living room. But that was exactly what Ehrenfest had been angling for as soon as he confirmed that both Niels Bohr and Albert Einstein would be in Holland to celebrate the long and fruitful career of Hendrik Lorentz, a man whom all three held in great esteem.

The occasion was the celebration of the fiftieth anniversary of Lorentz's doctorate, to be held on December 11, 1925. The Royal Netherlands Academy of Arts and Sciences would be honouring the dignified, white-bearded scientist with the creation of the Lorentz Medal, a gold medal to be awarded once every four years to a theoretical physicist in recognition of his important contributions to the discipline.[1]

Lorentz's celebration was bringing many of Europe's leading physicists—along with a few royals—to Holland, and Ehrenfest intended to take full advantage of the opportunity to see if he could do something to head off a burgeoning ideological clash between his two dearest friends. Bohr and Einstein had finally met at a conference in Berlin a few years earlier, and they'd taken quite a liking to each other, just as Ehrenfest had hoped. However, by 1925 it was clear to him that his two friends had parted company, ideologically speaking. Ehrenfest had set himself the daunting—and perhaps foolhardy—task of bringing the two men together to see if they could find some kind of common ground on the direction the confusing new physics was taking.

Ehrenfest had a professional motive for wanting to sit Einstein and Bohr down together. Quantum physics had become unutterably confusing, and he hoped that they could, together, find some path out of the tangle of conflicting ideas about how to proceed with the new science. Ehrenfest also had a personal motive for bringing his two dear friends together: simply, they were his friends. He didn't want them falling out and he didn't want ever to be in the position of having to chose between them. However cheeky it might seem for him to insist, like an admonishing parent, that two Nobel laureates sit themselves down and sort out their differences, if anyone could do it, it would be Ehrenfest.

When it came to getting Bohr and Einstein together in Leiden in 1925, Ehrenfest pulled out all the stops. Einstein, who revered Lorentz more than any other man on the planet, would not miss taking part in the festivities to celebrate his career. Einstein would definitely be there; getting Bohr to Holland was trickier. Bohr did not have quite the same attachment to Lorentz as Einstein and Ehrenfest, but once Ehrenfest assured him that Einstein would be there, Bohr accepted the invitation.

As the months of 1925 progressed, it became far more important to Ehrenfest to get the two scientists together than it had been in 1919. Back then, Ehrenfest had mainly wanted two men whom he really liked, and who shared his interest in the developments in physics, to meet and get to know each other. But so much had been happening just in 1925 alone that there was some urgency in keeping the growing schism between the two men from getting worse than it already was.

As soon as Bohr agreed to the trip to Leiden, Ehrenfest wrote to Einstein:

> Bohr is now struggling mightily with the problems of the quanta, and he needs to talk about his ideas with you more than anyone else. It is so important to him to know to what extent you have run into the same deep difficulties as he has. I know that no man alive has seen so deeply as you two into the real abysses of the quantum theory, and that no one but you two really sees what completely radical new concepts are needed.[2]

Bohr and Einstein were not, of course, the only people caught up in the tortuous struggle to make sense of a new science that just kept getting stranger and stranger. Sometimes it seemed as if the only way anyone was ever going to make sense of the atomic world was if they started throwing out some of the long-held beliefs that had provided such a sturdy foundation for physics for so many years. But what should stay and what should go was not at all clear. As two scientists with significant stature in the science community who were intimately familiar with the difficulties in quantum physics, Bohr and Einstein were both in a position to ease the growing friction between physicists who championed different views of what should stay or go. However, a public rift was developing between the two men on exactly that issue, since they had quite different ideas about what "radical new concepts" might be needed.

With so much at stake, Ehrenfest wanted the two men to go beyond what could be achieved in an afternoon discussion or a day of debate. He wanted them both at Leiden for a full week. He was offering the two men a retreat from the world, where they would be undisturbed by demands for their time and attention.

By this time, Einstein had become an international celebrity, and it wasn't easy for him to avoid media and public attention in whatever city he was visiting. Bohr wasn't a star like Einstein, but he was an honoured and much-admired man in Denmark, and he carried on his shoulders the burden of living up to his family's high standing in society and his position as a scientific leader with a burgeoning institute to run.

Einstein enjoyed his celebrity status most of the time. His rise to fame had come on the heels of a devastating war, at a time when the world was hungry for the kind of grand, sweeping ideas embodied in the theory of general relativity. It didn't matter if people had little idea what that theory was about. It was exciting and mysterious in its incomprehensibility, but most importantly, it had nothing to do with the ugly meat grinder of a war the world had just come through.

When the Royal Swedish Academy of Sciences announced in the fall of 1922 that Einstein had been awarded the 1921 Nobel Prize for Physics, suddenly the country Einstein had just temporarily fled because of the imminent threat of assassination wanted to claim him as a citizen. He was a Swiss citizen, but there was a bit of a political flap between the Swiss and German envoys over who got to claim the prize on his behalf. Since Einstein's initial acceptance of his position in Berlin tacitly implied he was a German citizen, the Germans won the honour of accepting the Nobel Prize, a triumph at a time when the country was suffering from postwar defeat and economic turmoil. Einstein was not particularly bothered by the fuss, as long as he officially retained his Swiss citizenship.

Einstein wore a number of hats, but his workload was hardly onerous. He held a paid position with the Prussian Academy of Science, with a joint appointment at the University of Berlin, and he was also the director of the Kaiser Wilhelm Institute for Physics in Berlin. His main task was to administer grants for physics research at German universities. He had no formal duties at the university, although he was one of its highest-paid professors. Neither did he have doctoral students to supervise. Einstein did some lecturing when he felt like it, and he served as the president of a foundation promoting experimental testing of general relativity. His only other professional obligation was his special professorial position in Leiden, and since that included an extended stay with the Ehrenfests once a year, it was not much of a hardship.

Bohr may have been a celebrity in Denmark, but he was hardly an international figure like Einstein. He did not have the worry (or pleasure) of being surrounded by giggling young women who hung on his every word whether they understood anything he said or not. Neither was he beset by reporters who wanted to know his opinion on social issues such as American prohibition and capital punishment (Bohr had not yet visited the United States), nor was he pestered by newsreel photographers to toss his hat in the air for the camera just one more time. Perhaps it was just as well. Bohr could hardly be oblivious to the fact that many physicists in Europe treated Einstein's fame—and his obvious enjoyment of it—with a certain derision and contempt, as if it were unseemly for any proper scientist to indulge in such pandering to the public.

In contrast to Einstein's relatively undemanding position in Germany, Bohr was practically a one-man physics show in his country. Denmark was not a big nation, but it had produced a few famous scientists between the sixteenth and nineteenth centuries, such as astronomer Tycho Brahe and Hans Christian Ørsted, the discoverer of electromagnetism. While his brother Harald was still a favoured son, he held the rather less prestigious position of a mathematics professor at the Polytechnic Institute in Copenhagen, so when it came to celebrity scientists in Denmark, Niels Bohr was it.

Bohr's working environment most certainly did not match his status in Danish society. His appointment as a professor at the University of Copenhagen in 1916, complete with a formal audience with the Danish king, officially inducted Bohr into the upper echelons of Danish society. But he had no laboratory for experiments and only a tiny 10-by-15-foot office at the university. The following year Bohr began the push to fund and build his own institute for theoretical physics. The new institute had to be approved by parliament, and that didn't happen until just before the war ended. Postwar inflation and political unrest stalled construction, and the costs kept going up. Due to strikes by masons and carpenters, the institute wasn't finished until the spring of 1921, at triple the initial estimate.[3]

Throughout the construction, Bohr worked closely with the architect on the building design and wrote up proposals for funding for everything from blackboards to chemicals. According to biographer Ruth Moore, he nearly drove his architect crazy:

> Whenever the architect finished a drawing, Bohr would see an opportunity to make an improvement. The plans would be redrawn. No sooner were they finished than another way to improve them would be revealed to Bohr. The architect alternated between collapse and explosion, but the plans were reworked and reworked until Bohr had the results he wanted.[4]

Such painstaking and laborious reworking and rewriting was Bohr's preferred working method, an approach he also took to his scientific tasks. Bohr, in his early thirties, was routinely seen racing from his house off-campus to the university on his bicycle or galloping up stairs two at a time. He was intense and hurried, with much to manage, "feeling vigorous and overworked,"[5] whilst vainly trying to keep up with the rapid developments in quantum physics, spurred mainly by the growing number of experimentalists devising new ways to test the behaviour of the quantized atom.

Before the new institute was even finished, young physicists began showing up, eager to work with Bohr, but he had nowhere to put them other than the adjacent library or a nearby laboratory. He had hired a secretary in 1919, but the office already housed Bohr and Kramers. There was no room for a third desk, so either the secretary or Kramers would stay home and let the other use the available desk, depending on which of them Bohr wanted to work with that day.[6]

The new institute was a simple but impressive three-storey building with a red-tiled roof that looked out over a park. It boasted a large lecture hall with terraced seating on the main floor and a library on the second, as well as many small offices and laboratory space. The top floor was intended as the Bohr family residence. By the time the institute was up and running, Bohr was exhausted. After the official opening in March 1921, Bohr postponed a series of lectures in Göttingen, let Ehrenfest present his partly finished paper at the Third Solvay Conference, and took a four-month break to recuperate.

One of the reasons Bohr needed to get away from the institute for a while was the unpleasantness that had been sparked by his growing opposition to Einstein's light quanta. The issue of light-is-a-particle/light-is-a-wave that Einstein had first raised in 1905 with his paper on the photoelectric effect—which he expanded on in 1909—had not garnered much attention, especially during the war years. Many scientists were too busy supporting the war effort in their particular country, and besides, the long-standing wave theory of light was quite satisfactory for the work they were doing.

Einstein, in his 1909 paper on light quanta, had derived a formula that included both a term for particles and a term for waves, so that there appeared to be a wavelike influence at work that couldn't be separated from light quanta. In 1911 he wrote to one of his friends that he'd stopped asking himself if light quanta actually existed. "Nor do I attempt any longer to construct them," he added, "since I know my brain is incapable of fathoming this problem."[7] From 1911 to 1915, Einstein had taken time out from working on the light quanta to complete the generalization of special relativity, but otherwise the puzzle of the light quanta was constantly on his mind.

The key problem with light quanta was that trying to combine the dual wave-and-particle nature of light was completely illogical. Think of a red rubber ball rolling across the floor. The ball is a specific entity that can be described by its colour, the material of which it's made, the geometry of a sphere and so on. At any given instant its position and momentum can be accurately predicted. You can even pick it up and hold it in your hand. A wave is quite different, because it is motion. Think of a rope tied to a fence post. If you grab the rope and swing it up and down, the rope will start to oscillate, creating a wave that will travel to the end of the rope and back again. The wave is not the rope or the fence post or your arm; it is the motion of the rope. Waves that are not confined to a rope, such as electromagnetic waves (radiation), form a field of continuous undulating motion. It is not something you can pick up and hold in your hand. Even if light quanta do not have the same properties as material objects like red rubber balls, the general idea is the same. Light, as both a wave and a particle, is hard to fathom because the characteristics of one preclude the other.

Theorists had long depended on experimentalists to validate or falsify their hypotheses, but experimentalists in the early 1900s were unable to resolve the wave/particle conflict. When they set up an experimental apparatus to test for waves, light obligingly behaved as a wave; when they set up an apparatus to test for particles, light obligingly behaved as if it consisted of particles. The experimental evidence served only to confirm that there was indeed a paradox.

Einstein understood the obvious incompatibility of waves and particles. In his 1909 paper he opined that physics must eventually come up with a theory of light that could be seen as "a kind of fusion" of the wave and the particle so that "wave structure and quantum structure . . . are not to be considered as mutually incompatible."[8] An explanation would surely emerge, eventually, to explain just how waves and particles influenced each other.

In 1916, when Einstein returned to work on the quantum, he went back to his earlier form of thinking about molecules and radiation. But this time he imagined what would happen if he combined

Boltzmann's container of gas molecules with Planck's container of electromagnetic waves. Einstein worked out, following Bohr, that when electrons jumped from one orbit to another, each transition resulted in the emission or absorption of a quantum of radiation. Furthermore, he reasoned, a light quantum, being an object with momentum, would impart some momentum to the molecule as it "hit" the molecule and was absorbed; and he built a wave-and-particle picture of how molecules interacted with radiation that he found quite satisfying.

Just as Rutherford had asked Bohr how electrons knew which orbit they were jumping to, Einstein asked himself how a light quantum could know in which direction to move. There seemed to be no causal explanation for either. "I do not doubt anymore the *reality* of radiation quanta," he wrote to his long-time friend Michele Besso, "although I still stand quite alone in this conviction."9

While Bohr's rejection of the light quanta was growing stronger, Einstein no longer doubted that they were real, even if he didn't have all the answers. But there was yet another problem.

Physicists, Bohr included, had long been uncomfortable with quantization because it so clearly violated the continuity that was fundamental to classical physics. The quantized Bohr atom was based on electrons being limited to the allowed orbits, which meant, in theory, that they couldn't be anywhere in between the orbits. So how did the electrons get from one orbit to another without passing through the space—albeit a very, very tiny space—between them? In the everyday world, an object could not appear in one place and then suddenly appear in another place without physically going *from* one place *to* the other.

Instantaneous transitions are forbidden in the world of classical physics, but if that's what was really happening at the atomic level, the atom was obviously following a different set of rules than those of classical physics. Experimentalists showed that electrons making a transition from one orbit to another emitted radiation exactly as predicted by the theory; the difference in the energy between the two orbits was precisely the emitted radiation frequency multiplied by Planck's h. But if an electron was physically travelling from one orbit to another, it would have to expend some of its energy to get there. That did not, however, appear to be the case.

For some physicists, the troubling "quantum jump" could be done away with simply by allowing that conservation of energy did not apply at the microscopic level in the same way it did at the macroscopic level. The electron really was travelling from one orbit to another rather than making an instantaneous jump, but it wasn't using up any energy in the process. This, however, would be a clear violation of the conservation of energy law, which expressly forbids work being done without expending some energy (and is the reason perpetual motion machines cannot exist).

Einstein had pondered that particular problem early on in his work on light quanta, deciding that one would have to give up the quantum so that the conservation law remained valid in all circumstances. But he couldn't do it. Even before he firmed up his belief in light quanta, he wasn't about to sacrifice them to save one of the key principles of the first law of thermodynamics. Who, he wondered, would have the courage to make such a decision?[10]

By the 1920s, many physicists working on quantum theory had bought into the idea that they could either accept that light was quantized or they could accept that energy was conserved, but they couldn't have both. Bohr had initially fallen firmly into the camp of quantum believers. It was, after all, the basis of his quantized atom and crucial to his 1913 trilogy. At first he accepted Einstein's theory of light quanta, but as time passed he started giving serious thought to the idea of conservation of energy as a statistical probability rather than an absolute law.

Bohr had been working for some time on what he called the correspondence principle, an attempt to reconcile his quantized atom with the rules for both the microscopic and macroscopic worlds. Electron orbits, according to his theory, got closer and closer together the farther they were from the nucleus. While the orbits closer to the nucleus required significant "jumps" of energy to get from one orbit to the next, orbits farther from the nucleus were so close together that the difference in energy to go from one to the other seemed almost continuous. Thus the microscopic world "corresponded" to the real world for the orbits farthest from the nucleus—as if Planck's h were not there anymore.

Bohr now had supporting experimental evidence to show that the electron orbits farther away from the nucleus were so close together that the electron transitions from one orbit to another were essentially continuous, and thus the physics of the larger orbits in the atomic world corresponded to the physics of the everyday world. He could clearly see that "every description of natural processes must be based on ideas which have been introduced and defined by the classical theory."[11]

But that presented a dilemma for Bohr. He had been thinking for some time that some of the quantum phenomena they were struggling to understand could be explained by letting go of the classical understanding of the behaviour of particles (or systems of particles) during interactions. He could retain classical conservation of energy and momentum by giving up a classical space-time description for particles, allowing that they had some kind of non-classical interdependence; or, he could retain the classical space-time picture—which did not allow for Einstein's light quanta, with their wave/particle interdependence—and let go of conservation of energy and momentum on average. So that's what he did.

Thus, when Bohr and Einstein were finally introduced by Max Planck in the spring of 1920 at a conference in Berlin, there was lots to discuss. Meeting Einstein was a big deal for Bohr. According to science historian Martin J. Klein, "The prospect of meeting with Einstein would have seemed even more momentous [than meeting J.J. Thomson] in Bohr's mind, particularly after all he had heard from Ehrenfest. Though Einstein was only six years older than he, Bohr would always see him as one of the grand masters, as if he were of another generation."[12]

Einstein and Bohr got into an enthusiastic debate within moments of being introduced. The two men were quite taken with each other. In his first letter to Einstein, Bohr said, "You don't know how very stimulating it was for me to have the long awaited opportunity to hear directly your views on the questions I have been working on."[13] Einstein was equally impressed with Bohr. At this point they were still on a somewhat similar scientific wavelength; but no matter how impressed Bohr was by Einstein, Bohr had not changed his mind about light quanta.

On the other hand, Hendrik Kramers, Bohr's assistant, quite liked the idea of light quanta. The anxious, bespectacled young Dutchman had graduated from the University of Leiden in the spring of 1919—the occasion when Bohr and Ehrenfest first met—and had then become Bohr's first research assistant. They worked well together. Bohr would come up with new and different approaches to the problems faced in matching a theory of the atom with experimental evidence; Kramers would take dictation from Bohr and provide the mathematics necessary to support Bohr's ideas. But they clashed over light quanta.

In 1921, Kramers took a look at some of the work by experimentalists such as the American Arthur Compton, who had been trying to figure out the arrangement of electrons in atoms by firing X-rays at them. Kramers worked out that if light quanta were real, they would lose some energy when they collided with an electron, and if they did, then the law of conservation of energy would hold. Kramers was very excited about what he'd come up with and took it to Bohr. But Bohr, burning out from the burden of his many responsibilities, did not take kindly to an open challenge to his stand against light quanta. According to Kramers's biographer Max Dresden, "Bohr and Kramers immediately started a series of daily no holds barred arguments."[14]

Kramers, who was often torn by doubts about his abilities as a scientist—the same dark doubts that plagued Ehrenfest—ended up depressed and exhausted, and finally in the hospital. For a man for whom "fear and anxiety about his role in physics were . . . constant companions,"[15] standing his ground against an adamant and unrelenting Bohr may very well have been impossible. "During his stay in hospital," wrote Dresden, "Kramers gave up the [light quanta] notion . . . altogether. Instead he soon became violently opposed to the [light quanta] notions, and never let an opportunity pass by to criticize or even ridicule the concept."[16]

The whole episode was hard on Bohr, and it seemed to have a negative effect on the other researchers working at the institute as well. It was a salutary lesson for young physicists at the beginning of their careers on the consequences of crossing Bohr. As Bohr wrote to Ehrenfest at the time, "You have no idea how much your friendship means to me. Specially at a time when I almost feel as a criminal in relation to all kinds of people here and elsewhere."[17]

But Bohr didn't relent, and even in his Nobel Prize lecture in December 1922 he dismissed light quanta as doing little to shed any light on the nature of radiation. At the time he had just finished writing a paper that cast the hypothesis of light quanta as merely a formalism. Yes, the existence of light quanta certainly provided a tidy explanation of the photoelectric effect, but it was quite unsatisfactory when applied to the quantized atom. According to Bohr, light quanta gave rise to "insuperable difficulties when applied to the explanation of the phenomena of (wave) interference" and excluded the possibility of a rational explanation of the role of frequency.[18]

Perhaps Kramers simply felt he was on safer ground as an ardent defender of Bohr's ideas than by sticking his neck out, espousing ideas of his own and suffering the pain of rejection. Kramers never published the work he took to Bohr. Then, in early 1923, Compton and a German experimentalist separately published the same idea that Kramers had come up with in 1921, with supporting experimental evidence. The Compton effect, as it was called, demonstrated clearly "first, that a light beam behaves as a bundle of light quanta, secondly, that energy and momentum are conserved" when light quanta and electrons collide.[19] That should have been enough to settle the issue of energy conservation.

Despite the experimental evidence for the particle nature of light, Kramers did not revert to his earlier thinking. Bohr didn't back down either. Instead, they continued to distance themselves from light quanta. In a 1923 paper Kramers said, "The theory of Light Quanta might be compared with a medicine which will cause the disease to vanish, but kills the patient. The fact must be emphasized that this theory in no way has sprung from Bohr theory . . ."[20]

In the summer of 1923, Einstein returned to Copenhagen for a second visit. (His first had been a short stopover on his way home from Norway in 1920.) Niels, Margrethe and their three boys were staying with Bohr's mother in her large apartment. Their own apartment on the top floor of the institute was still not ready to move into, but Bohr was already at work raising money to expand the institute. It would eventually include a new laboratory on one side and a three-storey house for the family on the other.

Bohr went to meet Einstein at the train station. He later recalled:

> We took the street car from the station and talked so animatedly about things that we went much too far past our destination. So we got off and went back. Thereafter we again went too far, I can't remember how many stops . . . in any case, we went back and forth many times in the street car and what people thought of us, that is something else.[21]

Obviously, the two men still had lots to talk about, and although they differed in their approach to the problems of making sense of the quantized atomic world, their differences had yet to harden into a public dispute.

Bohr's opposition to light quanta, however, was becoming more and more entrenched, something the young American physicist John Slater found out when he arrived in Copenhagen that winter.

Slater, with the ink barely dry on his doctoral degree from Harvard, had been trying to figure out some way to reconcile the wave and particle nature of light, just as Einstein was doing. He proposed a virtual field of waves, an idea similar to Einstein's "ghost field," which guided a light quantum to one of the allowed orbits and somehow permitted atoms in the field to communicate with each other during transitions. Further, he linked the continuity of the wave field with electrons settled in allowed orbits, while the discontinuity of quantization applied only when electrons jumped from one orbit to another and emitted light quanta. It was, he hoped, a nice tidy package that explained the paradoxes of light particles, waves, continuity and discontinuity.

Slater was quite unprepared for the reception he got from Bohr and Kramers. "I found . . . to my consternation, that they completely refused to admit the real existence of [light quanta]. It had never occurred to me that they would object to what seemed like so obvious a deduction from many types of experiments."[22] Slater, like Kramers, found it easier to bow to Bohr's opinion than to fight him, and decided that he could live without "the little lump carried along on the waves."[23]

Slater also didn't balk when Kramers and Bohr started drafting a paper to advance Bohr's proposal that energy was not conserved in atomic interactions and that light quanta did not exist. Slater hoped he'd get a chance to suggest changes as the paper was developed. He didn't. Bohr used Slater's virtual radiation field, not as part of the theory of light quanta (as Slater had originally intended), but as the basis for a new theory of radiation that had no place for light quanta at all.[24] Although the paper had Slater's name on it, he was shut out of its development. Bohr, he said, had done all the dictating, Kramers the writing. Kramers's excellent math skills were not required, since the paper contained only a single, short equation.

The hastily assembled BKS (Bohr, Kramers, Slater) proposal, as it was called, was published in January 1924 and contained some obscurely presented but quite radical proposals. In particular,

Bohr's new formulation of quantum physics called for the abandonment of conservation of energy for individual transitions between orbits while retaining conservation on average over a large number of transitions.

The BKS proposal retained the ideas that had so successfully explained spectral lines, but it took a different tack to explain transitions. Back in 1913, Rutherford had asked Bohr how an electron would know where it would stop before it "jumped," and this time Bohr had an answer. According to the BKS proposal, the emission of light from a transition was induced by probabilistic virtual fields, and hence "the atom is under no necessity of knowing what transition it is going to make ahead of time."[25] Attributing the emission of radiation during a transition to the influence of a virtual field that, unlike the electromagnetic field of radiation, wasn't actually there was just another way of saying that emissions had no real cause. Bohr was boldly abandoning two long-standing principles of classical physics—conservation of energy and causality. No quantum jumps, no causality and no conservation of energy—and no light quanta.

The reaction to the BKS paper in early 1924 was muted. Bohr's writing was obscure and difficult to understand, and there was no mathematical structure to help the reader work through the dense prose. His writing style was hard to fathom because he tended to "qualify his remarks so that no shade of meaning, no subtle difference between situations, could be overlooked by the reader. Bohr seemed to fear oversimplification more than anything else, and exerted all his efforts to avoid giving the illusion of clarity."[26] Some physicists expected a thorough mathematical explanation to follow the written proposal, assuming that then they'd be able to understand what Bohr was getting at. Such an explanation never appeared.

Bohr had circumvented the question of light quanta in the atom, in spite of Compton's experimental evidence to the contrary. But, at the time, there was no experimental proof that energy was conserved at the atomic level or that atomic transitions couldn't be acausal. By the 1920s the European physics community had shifted away from the Machian positivism that had been so influential earlier in the century. After the war, societal views of the world had changed from an appreciation of cold, practical facts to an embrace of the intuitive and surreal. The BKS proposal thus arrived at a time when chucking out the old ways had become much more acceptable.

Viennese physicist Erwin Schrödinger was the first to officially support the BKS proposal. Schrödinger had never liked the idea of the discontinuous quantum jumps, which could not be visualized, and he could live with the idea that energy was conserved only statistically.[27] Göttingen physicist Max Born was also quite excited by the BKS proposal, although he'd only got a verbal report from his young assistant Werner Heisenberg, who had just spent a couple of weeks with Bohr at Copenhagen. "I am quite convinced that your new theory hits the truth," Born wrote enthusiastically to Bohr.[28] Bohr and Einstein had not spoken since the previous summer, and they did not frequently exchange letters the way they did with Ehrenfest. Bohr was anxious to hear what Einstein thought of the paper, but it took him a while to find out.

Einstein wrote an article for a Berlin newspaper in April 1924 about the dual nature of light that included both light quanta and waves. "We now have two theories of light," he wrote, "both indispensable, but, it must be admitted, without any logical connection between them, despite twenty years of colossal effort by theoretical physicists."[29] The next month Einstein gave a talk in Berlin about the BKS proposal and publicly outlined his many objections. He was, of course, opposed to the abandonment of the law of conservation of energy, the abolition of light quanta and doing away with causality.

Within the physics community, at least amongst those involved in quantum physics, the disagreement between Bohr and Einstein was not a secret. One of Einstein's friends wrote to him that he'd just been to see Bohr. "How strange it is that the two of you, in the field where weaker imaginations and powers of judgement have long ago withered, alone have remained and now stand against each other in deep opposition."[30]

Bohr could have got Einstein's opinion directly from the horse's mouth, so to speak, if it hadn't been for his busy schedule. Bohr made a quick visit to Göttingen in June to talk to Born about the BKS proposal and about how Born was using some of the concepts from the paper in his own work. Bohr just missed Einstein's arrival. Born and Einstein had been friends since before the war, and while Born's assistant Heisenberg was elated to meet the great man, he and Born were both disappointed to learn of Einstein's objections to the BKS proposal.

In the summer of 1924, Bohr was once again in a state of exhaustion. He and Margrethe decided they needed a holiday home, and they bought a quaint thatched house near the edge of a forest and not far from the sea. By this time the Bohr boys numbered five, the youngest only a few months old. It was a wonderful place for children to play and an idyllic setting for a man whose many burdens were wearing him out. It didn't mean he escaped his work altogether, but at least he was working in a soothing environment.

It wasn't until the fall that Bohr got chapter and verse on Einstein's objections to the BKS proposal. Hamburg physicist Wolfgang Pauli had spent some time at the Bohr institute as a visitor the previous year. He'd had a long talk with Einstein about their shared reservations about the BKS paper when their paths crossed at Göttingen.

About the same time, the Danish and German newspapers got hold of the story. Two of the world's leading physicists, both Nobel laureates, were in conflict over the nature of light and conservation of energy, and a much-anticipated experiment being conducted in Germany was about to settle the dispute. It was a clash of the scientific titans, and of course the public always lapped up anything the newspapers wrote about Einstein. When a journalist from a Berlin newspaper contacted Einstein to get his side of the story, the media-savvy star would say only that there was indeed a difference between him and Bohr about the nature of light, but that there had not been any exchange of views about the subject between them.[31] He wasn't about to stoke the flames in the press.

Nonetheless, the dispute between Bohr and Einstein was now in the public domain, with sides clearly drawn. The standoff ended abruptly in the spring of 1925 when the German experimenters announced their results, which were quite unambiguous. "The results," the experimentalists wrote in their published paper, "do not appear to be reconcilable with the view of the statistical production of recoil and photoelectrons proposed by Bohr, Kramers and Slater. They are, on the other hand, in direct support of the view that energy and momentum are conserved during the interaction between radiation and individual electrons."[32]

Bohr was wrong, Einstein was right and the physics community in general was relieved that the conflict and confusion was over. Bohr accepted the experimental refutation with good grace, finding some comfort in knowing that the law of conservation of energy was no longer in doubt. Whatever their philosophical differences about how to proceed in quantum physics, Bohr still held Einstein in great esteem. But he was not going to give up his opposition to light quanta, even though he was fast becoming the last of the quanta deniers.

Pauli, who'd had his doubts about the proposal from the start, wrote to Kramers that "it was a magnificent stroke of luck that [BKS] was so rapidly refuted"[33] by solid experimental evidence. Slater, however, was bitter. While he hadn't been too bothered about giving up light quanta at Bohr's insistence, he had not wanted to give up conservation of energy. Now he found himself at a crucial stage in his early career with his name on a paper that was simply wrong—a paper he hadn't even been allowed to write.

It must have been music to Bohr's ears in September 1925 when Ehrenfest invited him to spend a week in quiet seclusion at Leiden.

Imagine, Ehrenfest wrote, quiet chats with Einstein while strolling along the canals near the house or taking long walks on nearby shores, without the worry of interruption. They could have the two small bedrooms upstairs, and Tatyana would even allow them to smoke in their rooms. Ehrenfest was quite insistent that Bohr not bring any writing work with him, knowing the Dane's tendency to laboriously and continuously rewrite his papers. "It would really be a sin to spoil this opportunity," Ehrenfest said, "this rare opportunity to peer for once with Einstein into the furthest deepest depths of physics that are at present accessible to anyone's gaze. And you would certainly, and at the same time uselessly, spoil it for yourself if you were to bring such work with you."[34]

Ehrenfest envisioned privacy, peace and quiet for his guests. But the turbulent times of 1925 were not yet over for Bohr.

Chapter Seven

TAKING A NEW PATH

When you come to a fork in the road, take it.

YOGI BERRA

Ehrenfest wasn't the only physicist feeling rather pleased with himself in the fall of 1925. Max Born was preparing to embark on a six-month tour of the U.S.A. to pitch the new Göttingen quantum mechanics, and his wife Hedi was coming with him. The American tour represented a remarkable shift in his scientific career in just a few short years, while Hedi's presence on the trip offered some hope that their relationship was on the mend.

Until 1921, Born had focused much of his scientific work either on Einstein's relativity or on the molecular structure of solid materials, but he'd always been interested in electron theory. He'd also had the responsibility of running his own theoretical physics institute at Frankfurt University, with its two rooms and chronic funding shortage. Like Bohr, he was continually burdened by his many administrative responsibilities, but he was also handicapped by breathing problems and severe asthma attacks that forced him to regularly seek the healing powers of spas and the clear mountain air of the Alps.

The big shift in Born's career accompanied his appointment as director of the Physical Institute at Göttingen in 1921. Göttingen was already the world leader in mathematics, and when Born arrived at his old alma mater he fully intended to turn the university into a world leader in the new atomic physics. He'd succeeded even though atomic physics was a new field for him, and he was quite pleased with the team of bright young men he'd brought to Göttingen to help build the institute's reputation.

Turning Göttingen into a leading physics institute was a mixed blessing for Born. Physics students, eager to learn from the best professors and make valuable career connections, flocked to the institute, as did young physicists and senior scientists. As the director, Born was responsible for meeting with and playing host to visiting scientists, and it all got to be a bit much at times. Wearied by the arrival of yet another couple of visiting physicists who wanted his attention, he once told Einstein, only half joking, that he was going to feign his death and simply refuse to see anyone so he could get some of his own work done.[1]

Einstein and Ehrenfest were both welcomed guests at the Born family home, and they made a point of spending a couple of weeks at Göttingen every summer for academic and social exchanges. Ehrenfest would usually arrive with an entourage, which often included Tatyana and their eldest daughter, who was studying physics and mathematics, along with a couple of assistants. Sometimes Ehrenfest even brought along his Ceylonese parrot, which he had taught to say, to the amusement of visiting scientists, "But, gentlemen, that is not physics."[2]

Born was a couple of years younger than Einstein. They were both German-born Jews with similar views on politics and a shared despair over the nationalistic madness that continued to grip Germany. But there had been a few ups and downs in their relationship over the years.

Albert usually arrived without his wife when he stayed with Max and Hedi. To put it mildly, Elsa and Hedi did not get along. In November 1920, barely a year after Elsa and Albert had married, Hedi took it upon herself to make it very clear to Elsa what she thought of her in a scathing six-page letter. She accused Elsa of everything from dulling Albert's judgment to taking advantage of him when he was too ill with his chronic stomach problems to defend himself.[3] The letter had, of course, a chilling effect on the relationship between Albert and Max, but two months later Albert wrote to Max to bury the hatchet, and Max was only too happy to agree.[4] Max, for his part, apologized to Albert for not preventing Hedi from writing such "hard and sharp words," even though he hadn't known what she was up to. "I know about only part of the unpleasant correspondence between our wives," Born told Einstein, "as my wife decided one day not to take me into her confidence any longer."[5]

Max was often baffled by Hedi's behaviour. He didn't really understand her, but then, he'd never really understood women in general. Hedi had an artistic temperament that was at odds with Max's logical and mathematical mind, and he despaired sometimes about how to reach her. Hedi had always been something of a hothouse flower, an indulged young woman. Perhaps, given Max's wealthy background and his considerable talent at the piano, she'd attributed to him richer social skills than he really possessed. He had been a handsome young man who cut a fine figure in white tie and tails, and he routinely mixed with the who's who of Breslau society at dinners, concerts and salons hosted by an aunt who had taken him under her wing after his father died. In reality, though, he was a very shy person with a delicate constitution, and he always had been.

Max's mother had died when he was only four years old, and his father, an embryonic biologist, had withdrawn into his own world. Max and his younger sister, Käthe, were left in the care of housekeepers and governesses, under the eagle eye of his autocratic grandmother, who would sweep in and out of their mansion delivering orders to the staff. The two children had their cousins to play with when they visited the various family estates and mansions, but Max's grandmother restricted his socializing due to his delicate health. He was a lonely little boy who grew up into a lonely young man.

University was good for Born. He had made no real friends during his gymnasium years, but in university he had bonded with other mathematics students, studied astronomy and learned the simple pleasures of raising a stein in a local tavern and rowing with friends on the river. Of course, such outdoor activities were sometimes aborted by his asthma attacks.

Eventually the time came when Born, like other German university students, needed to choose which leading professors he wanted to study with. Unlike in most other countries, the German and Austrian university systems did not offer undergraduate degrees. Students did not work through a specific curriculum, nor did they stay with a single institution. Rather, they were encouraged to visit a variety of universities, to take in lectures by professors whose work interested them, and eventually to attach themselves to a professor who would supervise their doctoral degree. There were no exams to write, either. The culmination of a student's work was the doctoral thesis and oral defence before an examining committee. Of course, to advance their careers, students benefited from attaching themselves to the most prestigious and highly-thought-of professors.

Born was leaning towards physics but then decided to do mathematics instead. The only sensible place for him to go was Göttingen, home of the great mathematical triumvirate of Felix Klein, David Hilbert and Hermann Minkowski. When Born arrived there in 1904, Göttingen was a small medieval town with a university, a far cry from the sophistication of Breslau that Born was accustomed to. But he was quickly absorbed into university life when his excellent penmanship and note-taking ability earned him the coveted position of David Hilbert's scribe.* Born's position also meant he could spend time with Hilbert clarifying the lecture notes, and thus get to know the professor. Teacher and student liked each other, and Max became a regular visitor to Hilbert's house.[6]

---

*It was a tradition at Göttingen that a selected student would transcribe the lectures given by a professor and provide copies for the university's mathematics reading room.

Minkowski and Hilbert were close friends, so, for Max, getting to know Hilbert meant getting to know Minkowski as well. The two men were considered a little unorthodox for the tradition-bound university. They picnicked, drank in beer halls and invited students to join them, crossing the usual social barrier between student and professor. The imperious Felix Klein, however, was a different kettle of fish. He was older and more demanding of his students, and his support could make careers, just as his animosity could break them. Unfortunately, Born had got on the wrong side of Klein early on when he unwittingly snubbed the senior scholar by declining to submit a thesis for a competition Klein wanted him to enter. Without Klein's support, Born knew his prospects of becoming a mathematician at Göttingen were not bright.[7]

Only a year after he arrived in Göttingen, Born became Hilbert's first assistant, a definite feather in his cap. It certainly appeared that he had the full backing of Hilbert and Minkowski for a doctorate in pure mathematics. But, having witnessed the mathematical genius of his mentors first-hand and lacking any real confidence in his own skills, Born decided instead to consider a more physics-oriented topic for his doctoral thesis.

Göttingen boasted the largest group of academics working on the new theory of the electron, and this field of research captured Born's imagination. However, because he felt he had to appease Klein if he was to have any hope of getting his degree, he chose to do his thesis on one of Klein's favourite practical applications of mathematics, the elasticity of solid materials. Born also wisely arranged to have his oral defence committee made up of scholars who were more sympathetic than the unforgiving Klein. His attempts to appease Klein were typically met with icy disapproval, so he often avoided Klein's lectures, one more thing for Klein to hold against him.

In the summer of 1906, Born presented his doctoral defence and graduated magna cum laude with a doctoral degree in mathematics. The problem was, he wasn't going to be a mathematician;

he was going to be a theoretical physicist, which was something of a gamble given that there were barely a half-dozen positions for theoretical physicists in all of Germany and Austria at the time. And after his unpleasant time at the university, Born vowed he would never again set foot in Göttingen.[8]

Following a brief stint to fulfill his mandatory military duty, Born had to decide where to do his habilitation. This was the necessary step between getting a doctoral degree and getting (one hoped) a position as a junior professor. Habilitation required doing original research under the supervision of a senior professor and then presenting the research in a lecture. But Born wasn't entirely sure what it meant to be a physicist, because most of his experience had been in mathematics. On the advice of one of his friends, he decided to go to Cambridge University to see how "real physicists" worked. Born made a brief and unfruitful visit to England to learn more about J.J. Thomson's model of the electron from the great man himself, but Thomson seemed to have little patience for young scientists from abroad.

Born then returned to the University of Breslau to do his habilitation. Casting about for a research topic, he sent a note to a researcher working at the Swiss patent office, requesting copies of the paper he had recently published on the principle of relativity. Einstein promptly sent them off to Born.

Unburdened by the necessity of earning a living, Born was able to work for two years on his own, combining Einstein's relativity with what he'd learned about the electrodynamics of the electron from the lectures given by Hilbert and Minkowski at Göttingen. According to Einstein's relativity, when an object moved at close to the speed of light, its mass grew, distance contracted and time expanded. This made describing the motion of a fast-moving electron complicated. Born came up with a better method for calculating the mass of an electron, but his considerable excitement at what he'd achieved was dashed when Minkowski published a very similar solution in the spring of 1908. With encouragement from friends and family, Born sent Minkowski a copy of his work, which resulted in Minkowski inviting him to return to Göttingen to do his habilitation under his supervision. So much for never again setting foot in Göttingen.

That November, Born was back in the same rooming house where he'd spent his student days. Two months later, the 44-year-old Minkowski died of a ruptured appendix.

Born was shaken by Minkowski's sudden death. A promising academic relationship with Minkowski was shattered before it had really begun, and Born had lost a mentor he greatly admired. Born was invited by the mathematics students to speak on their behalf at the funeral, and his tribute to Minkowski was so touching that it softened even Klein's heart.[9]

After the funeral, it fell to Born to sort through Minkowski's unfinished papers to see what could be completed. He succeeded in reconstructing and publishing only one paper.[10] It was, however, an important paper on Minkowski's idea of combining space and time into a four-dimensional continuum, an idea that Einstein would later put to good use in his generalized theory of relativity. Despite his lack of academic sponsor, Born kept working on the relativistic model of the electron, absorbing ideas published by Hendrik Lorentz. He published his first paper on relativity in 1909, concluding that electrons in the "plum pudding" model of the atom were arranged in rings.* Eventually, Born found a Göttingen theoretical physicist to replace Minkowski as the sponsor for his habilitation.

Born's article on relativity and the electron attracted the interest of Paul Ehrenfest, who had been a student at Göttingen at the same time as Born. Born sent Ehrenfest a copy of his paper and a lengthy letter, hoping to garner the fellow physicist's support. Ehrenfest,

---

*Born's 1909 paper was published several years before Ernest Rutherford proposed his orbital model of the atom.

however, spotted a flaw in Born's paper and published one of his own that presented a different view. Nonetheless, Born's work earned him his first invitation to speak at a gathering of physicists, a meeting held in Salzburg in September 1909 where Einstein was in attendance. Einstein was giving a talk, but not on relativity. His lecture was about light quanta. "It seems to me rather amusing," said Born. "Einstein had already proceeded beyond special relativity which he left to minor prophets, while he himself pondered the new riddles arising from the quantum structure of light."[11] Born, the minor prophet, met Einstein, the major prophet, for the first time at the Salzburg meeting.

Born finished his habilitation a month later, giving a lecture on his relativistic model of the atom, and became a proper Göttingen lecturer. He then took a break from his teaching duties in the summer of 1912 to give a course on relativity at the University of Chicago at the behest of Albert Michelson.* Michelson had visited Göttingen the year before, and the two men had become friends through their regular sets on the tennis court. While in Chicago, Born spent time in Michelson's lab studying his new methods for photographing atomic spectra, which got him thinking about what the spectra might be implying about the behaviour of atoms.

Born returned to Germany, and while still a shy man, he had now developed a gloss of sophistication. That fall he met the lovely and intelligent Hedi Ehrenberg, who had come to Göttingen to take some university courses. Max and Hedi had a proper and circumspect courtship, with Max making many train trips from Göttingen to Hedi's home in Leipzig, where her father was a law professor. Or he would travel to a nearby spa town where Hedi

---

*This was Born's first trip to the United States, and he made a point of spending time in New York City, visiting with some of his family's wealthy Jewish friends, as well as taking in such sights as the Grand Canyon and Niagara Falls. Before he returned to Göttingen to resume lecturing, he took a side trip on the Canadian Pacific transcontinental railway across the Prairies to Lake Louise.

went to restore herself in luxurious surroundings, as was the fashion amongst the young women of well-to-do families.

Hedi's mother, Helene, was a matriarch in the tradition of Max's autocratic grandmother, and she did not take kindly to Max's refusal to convert to Lutheranism so the couple could have a big church wedding. The outdoor wedding took place on August 2, 1913, in Berlin at Max's sister's home, followed by a dinner that was lavish enough to placate Helene.[12]

Born, like Bohr, took his work along with him on his honeymoon. He had with him Einstein's latest papers on the application of relativity to gravity, and scheduled their travels so they would be in Zurich to attend a lecture Einstein was giving. That was followed by a stop in Vienna, where Max was presenting a paper. The gathering of scientists in Vienna was abuzz with the papers recently published by the young Danish physicist Niels Bohr, on the quantized atom, in which electrons orbited a nucleus just as planets orbited the sun. Bohr had also made a strong argument that his model explained the atomic spectra that had so intrigued Born in Michelson's Chicago lab. Born was not particularly impressed by Bohr's efforts, however.

Unlike Margrethe Bohr, Hedi Born did not see herself as her husband's secretary and resented the time he spent attending to his science instead of her, especially on their honeymoon. Nonetheless, nine months later their first child was born. A few months after that, Max Planck offered Born a newly created professorial position in theoretical physics in Berlin. Max had finally achieved success in both his career and his personal life. Not long after their arrival in Berlin in 1915, Hedi gave birth to their second daughter.

Born considered his years in Berlin to be most remarkable, despite the "sorrows, excitements, privations and indignities" of living in Berlin during the war, because that's where he got to know Einstein. "It was the only period when I saw Einstein very frequently, at times almost daily, and when I could watch the working of his mind and learn his ideas on physics and on many other subjects."[13] They became good friends, and they would while away many an hour in the Borns' apartment with Max at the piano and Albert on violin. Max also had the pleasure of long discussions with Albert about his new generalized theory of relativity, a subject Max was getting to know very well.

But there was no escaping the realities of living in a country at war. Max was commandeered to work on sound-range equipment for the military, which sometimes meant inspecting equipment at the front. While he was away, Hedi struggled with a deep depression, and with food in short supply she scrambled to feed herself and their two little girls. Upon his return from the front, Max bundled the ailing Hedi off to a sanatorium, where she stayed for several months to regain her health.

The Borns survived the war, as did their families. But their wealth did not. Much of the family fortune that Max had grown up with was lost in worthless war bonds. Since Max and Hedi could no longer rely on family money, Max's career advancement took on a greater importance.

While Max was away on military business, Hedi had got word that he might be in line for the position of head of theoretical physics at Frankfurt University. She asked Einstein what he thought, and he replied, "Accept unconditionally . . . one should not refuse such an ideal post where one is completely independent."[14] Although Max was in a good position to replace Planck when he retired, it was not a sure thing, and Einstein warned Hedi against giving up the opportunity in Frankfurt for a future position that might not materialize.

Max accepted the position in Frankfurt, and he and Hedi moved there in the spring of 1919. The shortages of food and heating fuel continued to make life difficult for the family. By the fall, inflation was chewing up the budget for Max's two-room theoretical physics institute, and since Einstein was responsible for distributing research grants in Germany, Max turned to him for help. Einstein could promise only to see what he could do.

Einstein did indirectly provide a means for Born to make some money for his institute. Einstein suddenly became a star when the Eddington expedition confirmed his general theory of relativity. The public was enthralled by Einstein's new vision of the universe and was hungry to know more. In January 1920, Born organized three evening lectures on general relativity in a large auditorium at the university, charging admission. The lectures were a hit and Born was able to raise enough money to keep the institute going.

Einstein was impressed, and wrote to his friend that he admired his ability to save the institute from penury by giving lectures on relativity while still writing papers and looking after his family.[15] The fact was, however, that Hedi and the girls were all ill, and Max was struggling with his usual winter asthma.

As busy as Max was, he signed a book deal based on his lectures on relativity. And then, in addition to everything the Born family had on its plate, he was offered the position of chair of theoretical physics at Göttingen. This meant another move for the family at a time when Germany's transportation system was still in chaos, and there was, moreover, Hedi to consider. Would she appreciate leaving a city like Frankfurt for the cultural backwater of Göttingen?

Max finished writing his book, *Einstein's Theory of Relativity and Its Physical Foundations*, and sent the proofs to Einstein to read. He also sent Elsa the biographical note on Albert that he was including in the book. Max wanted to know if it struck the right tone and, above all, avoided "any suggestion of burning incense before the idol."[16]

Born's book caused a flap when it was released in the summer of 1920. Einstein was the focus of particularly virulent anti-Semitic attacks at the time, with relativity labelled a Jewish science and therefore highly suspect. But it was the inclusion of the biographical note and a photograph of Einstein that most horrified German scientists. Personal publicity and personal aggrandizement had no place in scientific literature. In their minds, Einstein was already thumbing his nose at their scientific code with his growing celebrity status. Despite, or because of, the controversy, the first printing of Born's book sold out quickly, but under pressure from his colleagues, and against the objections of his publisher, he removed the offending biographical note and photograph for the second printing.

Hedi had been sharply critical of how Einstein was handling the personal attacks from those who opposed his "Jewish" relativity, and she was not shy about sharing her blunt opinion with the great man himself. She would have preferred that he rise above the pettiness of his attackers, and she worried that Einstein didn't always hold his tongue when it would have been wiser to stay silent. This, of course, she blamed on Elsa's influence.

The cracks in the Born marriage were getting harder to ignore, and the pressures of the move to Göttingen weren't helping. Still, Max had high hopes for the move. As feared, it did not go particularly well. Just before they were to leave Frankfurt, thieves broke through the bars on a basement window of their house during the night and made away with a lot of silver, linens, two bicycles, and even the suit and shoes that Max had left downstairs.[17] Hedi was pregnant with their third child, and since their apartment in Göttingen was still being renovated, she took the girls and headed to Leipzig to stay with her father. By the early summer the family was reunited in Göttingen, and Hedi gave birth there to a baby boy, Gustav.

Unlike Einstein and Bohr, Born had not devoted much of his scientific career to quantum physics, but he had been following the developments since Hilbert and Minkowski had first introduced him to the puzzle of the electron. Born was nearly forty when he made the move to Göttingen. It's a widely quoted adage in mathematics and physics circles that a scientist does his best work before the age of thirty, and after that his time has passed. Born's career had been quite respectable, but Göttingen was an opportunity to begin again in a new and rapidly expanding area of physics, and with clout and financial resources he hadn't had at Frankfurt.

In Göttingen, as was the practice in Germany, a senior professor like Born got to run his own program. There was already a
physicist heading up general physics. Born would have his own
institute of atomic physics, with enough funding for two assistants. The university had also opened up another new position for
an experimentalist with his own paid assistants, and Born immediately negotiated to get that position for his friend James Franck,
who would provide the experimental component of Born's atomic
research. Göttingen would now boast the First Institute of Physics,
Born's Institute of Theoretical Physics and Franck's Second Institute of Physics.[18]

For this venture to get off the ground, however, Born needed
some talented people around him—people such as Wolfgang Pauli.
While still at Frankfurt, Born had been following the work of the
young man who was studying with Arnold Sommerfeld in Munich.
Pauli had published an encyclopedia article on general relativity
that so impressed Born that he invited the student to come to
Frankfurt to work with him. In particular, Born wanted to develop
a theoretical basis for the electron orbits of Bohr's quantized
atom, something Sommerfeld had been working on for years with
some limited success. In an effort to coax Pauli to Frankfurt, Born
pointed out that they shared some ideas about the puzzle of quantum jumps that defied the continuity of classical physics. "You
regard the application of the continuum theory to the interior of
the electron as meaningless," wrote Born, "because it is principally
not a question of observable things. I have pursued just these
thoughts for a long time . . ."[19] But Pauli turned down the offer,
preferring to stay in Munich to finish his doctoral degree.

Sommerfeld, who was also very impressed with Pauli, had asked
him to take over the writing of the encyclopedia article on relativity
for him. Born's old nemesis Felix Klein was still at it. The elderly
Klein had long since retired from teaching, but he was overseeing
the compilation of the *Mathematical Encyclopedia*, the same work to
which Paul and Tatyana Ehrenfest had contributed. Born too was

writing an article for the *Mathematical Encyclopedia*, one based on research he had been doing years ago on the structure of solids. Klein hadn't asked him to do it; that had been Sommerfeld, back in 1915. For Born, his article on the "Atomic Theory of the Solid State" had become a tiresome chore that dragged on endlessly.

In early 1921, Born wrote to Einstein: "Pauli's article for the encyclopedia is apparently finished, and the weight of the paper is said to be 2½ kilos. This should give some indication of its intellectual weight. The little chap is not only clever but industrious as well."[20]

Needing an assistant in Göttingen, Born once again invited Pauli to come and work with him. This time, Pauli said yes.

Chapter Eight

ONLY WHAT THE EYE CAN SEE

Seeing is not always believing.
MARTIN LUTHER KING JR.

Max Born and Wolfgang Pauli were very different people, both in personality and in their approaches to science. Born was a model of the good German scientist: prompt, neatly dressed, cognizant of his status in the science community, and an early riser. Not so Pauli. He did not particularly concern himself with what others might think of him, and he was a night owl who enthusiastically embraced the nightlife of the big city. He didn't much like getting up before noon. There was not, however, much nightlife to speak of in Göttingen, so Pauli would often stay up late working because there was nothing else to do. Born would sometimes find him in his office late at night "rocking slowly like a praying Buddha,"[1] a habit Pauli had developed when he was deep in thought.

When Born succumbed to yet another episode of bronchial troubles that winter, he needed Pauli to give his lectures, something Born considered Pauli quite capable of doing even though he was only twenty-one. But because the lectures were at eleven in the morning, if there was to be any hope of Pauli getting to the lectures on time, Born had to send a maid to Pauli's rooms to rouse him out of bed.[2]

Given that Born did his doctoral degree in mathematics, it's hardly surprising that he would approach physics problems with an eye to describing them mathematically. Pauli was certainly mathematically adept, but he preferred a more intuitive approach to problem solving. Still, heading to Göttingen had all the marks of a good career move for the young man. He'd trained under Arnold Sommerfeld, who had started working on problems with the quantized atom not long after Bohr had come up with the idea. Born was prepared to look at the quantized atom with the fresh eyes of a newcomer and without any preconceived ideas or biases based on what had gone before. That gave Pauli the opportunity to start testing out his own ideas, as much as he continued to admire Sommerfeld.

Pauli had spent only three years studying at the University of Munich under the supervision of Sommerfeld, the shortest time period the university allowed for a doctoral degree. He had been working on the encyclopedia article on relativity that Sommerfeld had handed off to him and his doctoral thesis at the same time, a very demanding workload for anyone. But he also had to eat and keep a roof over his head, which had become more challenging with soaring inflation. Pauli's father, a biochemist and professor at the University of Vienna, sent him money to supplement his modest university grant, but as often as not, by the time the distracted Pauli remembered to go to the bank to collect the money from his father, inflation had rendered it almost worthless.[3]

Pauli developed a reputation early for having an unapologetic lack of respect for authority figures and for being a harsh critic. While still a doctoral student in Munich, he attended a lecture given by Einstein and opined from the back of the hall, "You know, what Mr. Einstein said is not so stupid."[4] This cheeky comment predated Einstein's Nobel Prize, but that probably wouldn't have made any difference to Pauli.

Werner Heisenberg started university in Munich two years after Pauli, and he was soon introduced to the teacher's assistant who was grading his homework. Pauli's blunt assessment of Heisenberg—"you are a complete fool"[5]—spurred Heisenberg to work harder. Always the pragmatist, Heisenberg quickly learned the value of Pauli's critiques in making his own work better.

Pauli's unconventional and idiosyncratic personality was honestly come by. Not only was he descended from a long line of high achievers, he grew up under the bristle-browed eagle eye of the redoubtable Ernst Mach.

Pauli's enterprising great-grandfather, Wolf Pascheles, had started a prosperous publishing business in Prague in 1827 at the age of fourteen, a business that continued to flourish under the guidance of his descendants.[6] Pauli's father, also named Wolf Pascheles, and Ernst Mach's son Ludwig had been friends from the time they were schoolchildren, and they later went to medical school together. Wolf fell completely under the spell of the senior Mach, a man who was one of the most influential philosophers and scientists in the Austro-Hungarian Empire in the latter half of the 1800s.

Mach, an experimental physicist and philosopher, had embraced the positivist tenets of French philosopher Auguste Comte as the rules for determining what constituted legitimate scientific inquiry, adjusted to fit his own particular outlook. Comte had lived through the French Revolution's infamous Reign of Terror at the end of the eighteenth century as well as the rise and fall of Napoleonic France, and he had witnessed first-hand the brutal suppression of political radicalism in the streets of Paris. More than anything, Comte wanted to inspire people to rise above such vicious and bloody madness. But to do so, he believed it was crucial for society to shake off the poisonous bonds of theology and metaphysics—to let go of such destructive fictions as the divine powers of royalty, religion and God. He sought an egalitarian, educated society of order and rational dialogue, where truth and knowledge were "based upon an exact view of the real facts of the case."[7]

In the mid-1800s, Ernst Mach was alarmed by what he perceived as the decline of European society into mysticism and anti-science, and he turned to positivism to "save" science. He wanted to anchor science in something more concrete than the latest whims and fancies of philosophical posturing. For this, he refined Comte's positivist philosophy; the "real facts of the case" were to be determined solely by human sensory perception. The only meaningful statements a scientist could make were about what could be measured, counted, tasted or otherwise experienced by the senses. This view, obviously, precluded discussion of ideas or concepts beyond the experience of the senses.

Thus was Mach able to effectively strip science of abstractions, metaphysical concepts and appeals to the supernatural, as well as the need for theoretical physicists. Theorists had no meaningful role to play in experiment-based positivist science. By taking a hard line against anything that was not subject to sensory perception or—heaven forfend!—that invoked God, Mach spearheaded an influential scientific movement that made the rules for legitimate scientific inquiry quite clear and denounced those who deviated from those rules.

Mach was also concerned about the growing fragmentation in the study of psychology, physiology and the physical sciences, each with its own language and methodology, as if they had nothing in common. He embarked on his own version of a unified theory, based on the idea that once brought together, the sciences of mind, body and nature could (more or less) explain the whole of the human experience. "I only seek to adopt in physics," Mach wrote in *Analysis of Sensations* in 1886, "a point of view that need not be changed immediately on glancing over into the domain of another science; for, ultimately, all must form one whole."[8]

Mach was not alone in seeking a unified world picture. According to science historian Gerald Holton, at the beginning of the twentieth century, German literature was filled with "a seemingly obsessive flood of books and essays on the oneness of the world picture." Indeed, in 1912, thirty-four of Europe's noted academics and philosophers—including Sigmund Freud, David Hilbert, Ernst Mach and Albert Einstein—signed a manifesto calling on Europe's scholars to combine their efforts in bringing about a comprehensive, unified picture of the world.[9] Nothing came of the manifesto, but it was a measure of the importance of a unified world view in Europe. It was also one of the reasons that Mach's positivist approach to physics found fertile ground in the European physics community, particularly in France.

At the time, however, German society was more amenable to the philosophy of the eighteenth-century German philosopher Immanuel Kant than it was to Machian positivism.[10] Even so, both Kantian philosophy and positivism had their roots in the eighteenth-century Enlightenment. This philosophical movement celebrated the triumph of human reason and elevated the experience of self above the ancient regimes of religious and philosophical authorities. Kant and Mach were on the same wavelength in that respect, but Kant allowed that God existed. The Kantian philosophy had a great advantage in academic circles in Germany and Austria, because an official religious designation continued to be mandatory for anyone aspiring to the position of professor.

The dominant opposing ideology to positivism was realism. While scientific realists in the mid-1800s were just as keen on experimental verification as the positivists, they did not see the necessity of severely limiting legitimate scientific inquiry to only what could be observed or measured. Realists believe that a real, external world exists, whether there is anyone there to observe it or not; thus, theories about the existence of atoms in 1877—or parallel universes and extraterrestrial life forms today—are legitimate lines of inquiry.

A simple (and admittedly simplistic) way to determine whether a person leans towards positivism or realism is to pose the old paradoxical question: If a tree falls in the forest and there's no one there to hear it, does it still make a sound? A realist will emphatically reply, "Of course! Trees make noise when they fall whether anyone is there or not." Asked the same question, a positivist will look at you as if you're a bit daft and reply, "If there's no one there, how can anyone know if it makes a sound or not? The question is meaningless." A Kantian, on the other hand, will agree that no one can know if the tree makes a sound since there is no observer, but will allow that God will know.

Europe had experienced a number of pendulum swings between positivism and realism since the Enlightenment, with one being the dominant philosophy for about thirty or forty years before it was the other one's turn. Realism had been the dominant movement from the 1840s to the 1870s, producing the laws of thermodynamics and conservation of energy, electromagnetic theory, the periodic table of elements, and, in other fields, the cell theory in biology and evolution by natural selection.[11] But when realists started proclaiming arrogantly that all natural phenomena could be reduced to the sterility of mere physics and chemistry, they triggered a backlash. That's when Mach's positivism came to the fore, advancing a friendlier, human-centred ideology based on the appeal to human reason and rationality.

But the Machians, according to historian of science Stephen G. Brush, eventually fell prey to hubris as well, with their strict reliance on experimental verification or empiricism. "In fact," noted Brush, "they went too far in the direction of blind empiricism, ridiculing all attempts to discover the real nature of unobservable entities. 'Positivism,' perhaps more accurately called 'negativism,' had all but killed theoretical physics in France; it now spread to the rest of Europe."[12] The return of realism came on the heels of what was ultimately, said Brush, the failure of one of the great projects of nineteenth-century physics: "to construct a mechanical or atomic model of matter and of the aether which would enable one to explain thermal and electromagnetic properties by means of Newtonian mechanics."[13] That failure led directly to the development of Boltzmann's probabilistic physics and Planck's quantum.

To realists who rebelled at the positivist restrictions on what constituted legitimate scientific inquiry, positivist mathematician Henri Poincaré said in 1905: "To those who feel we have gone too far in our limitations on the domain accessible to scientists, I reply: those questions which we forbid you to investigate and which you so regret, are not only insoluble, they are illusory and devoid of meaning."[14]

The physics community in Germany and Austria was not blind

to the particular scorn Mach and his followers reserved for the theories of Ludwig Boltzmann, who dared to use unobservable atoms and the mathematics of gambling in his theories. But scientists such as Einstein were asking those "illusory" questions in their research anyway. All five of his 1905 papers were based on concepts such as atoms and light quanta that were considered by positivists to be unobservable and hence meaningless.

Perhaps one of Mach's most influential physics books was *The Development of Mechanics*, published in 1883. Einstein had read Mach's *Mechanics* when he was still a student at ETH, a book he said had a profound influence on him. Einstein later noted that "even those who consider themselves as opponents of Mach are hardly aware of how much of Mach's way of thinking they imbibed, so to speak, with their mother's milk."[15]

Mach continued to publish books on positivism, often with help from Wolf, a.k.a. Wolfgang Pauli. After Wolf's father died in 1897, he had taken the not uncommon step of changing his name and religion in an effort to advance his academic career. He converted to Catholicism and chose the Germanized name of Wolfgang Pauli, and then moved from Prague to Vienna to take an academic position. Wolfgang married Bertha Schütz in 1899, and son Wolfgang Jr. was born the following year.[16] Mach, despite his strong anti-religious stance, agreed to be the child's godfather.

Mach's retirement didn't slow him down. He continued to provoke and irritate the science community, but also drew the ire of non-scientists such as Vladimir Lenin. There were a lot of similarities between Mach's egalitarian positivism and Lenin's own brand of Marxist philosophy; therefore, Lenin felt it necessary to go to considerable lengths to distance himself from Mach's philosophy by condemning "Mach and Co." in his 1908 polemic *Materialism and Empirio-Criticism*.[17]

As combative and harshly critical as Mach might be towards those who didn't recognize the obvious correctness of his position, he was charming, engaging and benevolent to those he cared for, including his godson.[18] Little Wolfgang Pauli Jr. did indeed imbibe positivist philosophy with his mother's milk. His father was very much under the spell of Mach, and from the time he was in diapers Wolfgang was immersed in Machian philosophy. Young Pauli would visit Mach's apartment, which he later recalled "teemed with prisms, spectroscopes, stroboscopes, electrostatic generators, and other machines."[19] The elderly, bearded scientist, his movements limited by partial paralysis, would explain his latest experimental ideas to the boy. When Pauli turned thirteen, his gift from his godfather was a copy of *The Development of Mechanics*. It was apparent from an early age that Pauli was destined for a career in physics.

Mach died when Pauli was sixteen, and by then Mach's influence had begun to fade. His adamant refusal to accept the existence of atoms, even with experimental evidence, made him look like a foolish old man. However, not long after Mach's death, his philosophy was resurrected by the Vienna Circle. The Circle was a collection of mainly mathematicians, physicists, economists and sociologists whose aim was to refine Machian positivism. They added another dimension to Mach's philosophy by allowing logic, as well as observables, to be an arbiter of truth. And since mathematics was firmly grounded in axiomatic logic, mathematics could be an arbiter of truth too.

The rise of realism had been truncated by the war, so the re-emergence of positivism as advocated by the Vienna Circle overlapped realism for a while, making it confusing to figure out which philosophy was holding sway. In many ways, the differentiation between positivism, logical positivism, realism and Kantianism in postwar Germany had much to do with the acceptance or rejection of the existence of God and/or an external world that exists whether or not humans are around to experience it. In brief: positivism has no God and no external world; logical positivism has no God and no external world but does have mathematical logic: Kantianism has no external world but does have God; and realism allows for both God and an external world.

Paul Ehrenfest was well acquainted with some of the founding members of the Vienna Circle. Two of them, Philipp Frank and Hans Hahn, were Ehrenfest's pals from their student days at the University of Vienna. Ehrenfest, Hahn, Gustav Herglotz and another student had been fast friends who called themselves the "inseparable four."[20] Ironically, positivists Hahn and Frank both had Machian arch-enemy Boltzmann as their supervisor, as did Ehrenfest and Herglotz. Frank, who was a couple of years younger than the others, was still working on his doctoral degree under Boltzmann when Boltzmann committed suicide.

The "inseparable four" and the founders of the nascent Vienna Circle scattered after graduation and habilitation. Hahn went off to teach at a remote Austrian university, returning to Vienna in 1921, while Ehrenfest went off to St. Petersburg and then landed the job in Leiden in 1912. Frank attracted the attention of Einstein in 1907 with a paper he published on causality, and the two men became good friends, both having a strong interest in the philosophy of science. When Einstein decided to leave Prague for Berlin in 1912 and couldn't persuade Ehrenfest to declare an official religion, he recommended Frank to replace him as professor of theoretical physics, and there Frank stayed. When Hahn finally returned to the University of Vienna in 1921 as professor of mathematics, the Vienna Circle got down to the serious business of promoting logical positivism.

Herglotz stayed in Göttingen teaching for a while before being offered a full professorship at the University of Leipzig. He stayed there until 1925, when a position became vacant in Göttingen, where his new home was just a few minutes' walk away from the apartment where Max and Hedi Born lived.

Pauli was not an ardent positivist, despite his upbringing, but he had little time for metaphysics and religion. He thought it rather funny that he'd been baptized as "Anti-metaphysical"[21] instead of Roman Catholic. By a priest. In a Catholic church. Only Mach could have pulled off a cheeky stunt like that.

Pauli Sr. was a well-known biochemist in Vienna, and Wolfgang's mother, Bertha, was a journalist writing for one of Vienna's most influential newspapers. Being a female journalist in the patriarchal culture of Austria was no small feat, but Bertha was prone to bouts of harsh self-criticism and suffered from anxiety attacks that could strike without warning.[22]

Wolfgang was not subject to his mother's debilitating anxiety attacks, but, like her, he would sometimes sink into a dark depression when, by his own judgment, he failed to fulfill the high expectations he had for himself. His godfather's influence could be seen in Pauli's sharp tongue and biting sarcasm, as well as his unrepentant disrespect for authority figures, but absolutely none of Mach's experimental acumen rubbed off on Pauli.

After only a semester working with Born in Göttingen, Pauli had had enough of living in a rural backwater and headed to Hamburg in early 1922 to complete his habilitation there. It wasn't long before it became apparent to those around him that the portly Pauli and experimental equipment were antithetical. It would be an understatement to say that Pauli was not a particularly graceful or athletic person; in fact, he was awkward, not very good with his hands and a disaster in the lab. It was the beginning of what came to be called the Pauli Effect. Said physicist Rudolph Peierls, a contemporary of Pauli's:

> This was a kind of spell he was supposed to cast on people and objects in his [vicinity], particularly in physics laboratories, causing accidents of all sorts. Machines would stop running when he arrived in a laboratory, a glass apparatus would suddenly break, a leak would appear in a vacuum system, but none of these accidents would ever hurt or inconvenience Pauli himself.[23]

Pauli thought it all rather funny, but one Hamburg experimentalist forbade him ever to set foot in his lab, and would consult with Pauli only through the lab's closed door.[24]

In the summer of 1922, Pauli returned to Göttingen to attend a series of lectures by Niels Bohr, an event organized by Born and Hilbert. The "Bohr Fest" marked Göttingen's arrival as one of the world centres for research in atomic physics, joining Copenhagen and Munich. Bohr packed the lecture hall with the who's who of physics circles, but there was room at the back for students such as Werner Heisenberg and Pascual Jordan. Bohr took time to chat with Heisenberg and Pauli afterwards, and ended up inviting them both to Copenhagen. Heisenberg was already committed to study in Göttingen with Born while Sommerfeld was away on an American lecture tour, so he couldn't go. But Pauli quickly arranged leave from Hamburg and showed up at Bohr's new institute in the fall. Here was a scientist whose intuitive approach to physics was a better match for Pauli than Born's mathematical inclination. "A new phase of my scientific life began when I met Niels Bohr personally for the first time," said Pauli.[25]

When Ehrenfest and Pauli met for the first time at that same meeting in Göttingen, Pauli was rather rude. "I like your publications better than you," Ehrenfest retorted.

"Strange," said Pauli, "my feeling about you is just the opposite."

Ehrenfest, who quite enjoyed the cut and thrust of such repartee, took it all in good humour, and the two thereafter routinely engaged in bouts of enthusiastic verbal jousting.[26]

Pauli's time in Copenhagen proved to be more frustrating than fruitful, although he found compatible scientific spirits in Bohr and Kramers. He returned to Hamburg and completed his habilitation in January 1924, but he didn't make his first scientific breakthrough until December.

Since no one could actually see an atom or electron or nucleus, their behaviour could only be deduced from the available experimental evidence. But in 1924, Bohr's quantum rules, along with some generalizations from Sommerfeld and other scientists, still did not answer all the questions raised by experimental evidence. In particular, there was still no way to explain—without invoking a little mathematical jiggery-pokery—why electrons arranged themselves in certain numbers in certain allowed orbits (shells). The orbit closest to the nucleus could have only 2 electrons; the second orbit, 8; the third, 18, and so on.

Electrons were already assigned three quantum numbers (n, k, m) to describe their state in an atom. Pauli hit on the idea that perhaps electrons had a kind of two-handedness, which meant adding a fourth quantum number (m_2) that could have two values. He then proposed that a key characteristic of electrons was that they were very exclusive: an atom could not have more than one electron with the same four quantum numbers. By counting off the allowed values for the quantum, according to the restrictions he'd imposed, he came up with 2, 8, 18, 32 and so on, matching the arrangement of electrons in their shells.

Pauli's exclusion principle did not bring him a lot of satisfaction because it was incompatible with Bohr's orbital model. Bohr and Pauli joked, with no small measure of frustration, about the "nonsense" and "swindle" of the current theory of quantum physics, with its untenable mixture of quantum and classical physics used to manipulate a result so as to match experimental evidence. By the spring of 1925, Pauli had had enough. Yes, he'd had a breakthrough with the electron exclusiveness, but for him it just made the whole quantum muddle worse than ever. In despair, he was ready to quit physics altogether. "Physics is at the moment very wrong," he wrote to a friend in May. "For me in any case it is much too difficult, and I wish that I were a film comedian or something similar and had never heard of physics!"[27]

But rescue was at hand. Heisenberg, recovering from a debilitating bout of allergies, had taken a curious mathematical leap in the dark.

Chapter Nine

THE EMERGNCE OF THE BOYS' CLUB

It is only an illusion that youth is happy,
an illusion of those who have lost it.

W. SOMERSET MAUGHAM

When Werner Heisenberg fled Göttingen in early June 1925, he could barely function. His plant and pollen allergies were dreadful, and Max Born, who was no stranger to breathing troubles, quite understood his need to get away to the rocky island of Helgoland. For the first few days the 23-year-old Heisenberg could barely even think, but after a while his face lost its frightening red puffiness, and before long he could once again muster his brain cells to tackle the puzzle of the quantum.

When he did, he came up with yet another model for the quantized atom. This time he reversed the usual approach to describing the frustratingly elusive properties of the Bohr atom. It occurred to him that it really made no sense to struggle so hard to describe electron orbits and how electrons jumped—or didn't jump—when no one could tell what the electrons were actually doing. He, Born and Pascual Jordan in Göttingen, Wolfgang Pauli in Hamburg, Arnold Sommerfeld in Munich, and Niels Bohr and Hendrik Kramers in Copenhagen had been hung up (in varying degrees) on trying to make the experimental evidence, especially the fine details of spectral analysis, match up with some kind of mechanical description of electron orbits. Granted, they did have some of the pieces to the puzzle, but not enough to create any kind of sensible picture of the wretched atom.

What if, Heisenberg wondered, they all just quit trying to make up some kind of description for things they couldn't see and had no way of testing? What if they limited the description of the quantized atom to only that which could be seen? Experimental results could be seen, but electron orbits, if electrons even orbited, could not.

It was just an idea, a possibility, but Heisenberg made a point of stopping in Hamburg on his way back to Göttingen to see what Pauli thought of it. He'd come to rely on Pauli to rein him in when his intuitive leaps launched him in directions that couldn't be supported by the physics. Indeed, Pauli had warned Bohr about Heisenberg's sudden infatuations with this or that idea just before Heisenberg's first trip to Copenhagen. "Things go very oddly with him," Pauli told Bohr. "When I think about his ideas, they strike me as dreadful and I curse myself over them. He is quite unphilosophical—he pays no attention to working out clear principles or connecting them to existing theory."[1]

After Pauli's reaction to the last model he had proposed, Heisenberg was being cautious, but this time Pauli was not outright dismissive. Maybe Heisenberg was on to something, but it was just the beginning of an idea that needed a great deal more work to put a proper foundation under it.

When Heisenberg got back to Göttingen, he was ready to see where this new approach might take him. Born and Jordan were busy working on a mathematical model of the atom that didn't interest Heisenberg, so he quietly worked away on how to describe atomic behaviour using only observables, such as energy and the intensity and frequency of emissions, that could be experimentally tested by spectral analysis. It took him into tortuous mathematical territory, but rather than ask advice of Born or Pauli, he kept working away by himself. He was being a bit more careful about sticking his neck out on a half-baked theory.

The significance of observables in physics was hardly new. The philosophy of Mach placed great emphasis on the use of only those elements in physics that could be measured, counted or otherwise observed. The Vienna Circle had resurrected interest in positivist philosophy, and the group—originally called the Ernst Mach Society—expanded its philosophy of science to include an atomic theory that embraced the ideas of Ludwig Boltzmann (which must have had Mach spinning in his proverbial grave) and Albert Einstein, along with David Hilbert's pronouncements on mathematical logic.[2]

If anybody knew about the significance of limiting theories to observables, it was Pauli. But none of the Göttingen or Copenhagen crews evinced any particular interest in positivist philosophy. As Vienna Circle founder Philipp Frank discovered, German mathematicians and physicists didn't usually demonstrate overt philosophical leanings, at least on the job. "Most of them," said Frank, "did not try to bring their scientific activity into logical connection with their philosophy. They regarded the former as a thing of the intellect, the latter as a thing of the heart and the spirit."[3]

Still, as Einstein observed, scientists may very well have absorbed far more of Mach's thinking than they realized, even if they had no particularly strong feelings about any philosophy of science. Observables were in the air, so to speak, in both Göttingen and Copenhagen. In April 1925, just after Heisenberg had returned to Göttingen after another stint in Copenhagen, and before his attack of hay fever, Born and Jordan published a paper that emphasized quite clearly the importance of observability. "A postulate of great reach and fruitfulness," they wrote, "states that only such quantities should enter into true laws of nature which are in principle observable and measurable."[4] Kramers, Bohr's assistant in Copenhagen, had also made a point of using only observables in a joint paper he and Heisenberg had produced in a testy and awkward collaboration the previous winter, during Heisenberg's extended stay at Copenhagen.

Kramers and Heisenberg struck sparks off each other right from the start. Kramers was very protective of his status as the second-in-command to Bohr. If Bohr was the "Pope" at the Institute for Theoretical Physics in Copenhagen, Kramers was "His Eminence,"[5] a title that Pauli and Heisenberg used with more mockery than flattery. Kramers suffered from insecurity, and he was keenly aware that Bohr was surrounded by ambitious young wunderkinder—just like Heisenberg—who might displace him as second-in-command if given half a chance. Kramers recognized Heisenberg as a serious competitor for the position of Bohr's favoured assistant, but Kramers had the competitive advantage: he was four years older, and he was already an insider in the charmed circle around Bohr.[6]

For Heisenberg, Kramers's prickliness may very well have been a reminder of his relationship with his only brother. Heisenberg's father had deliberately provoked competition between him and his brother, Erwin, from the time they were little boys. Erwin was a year older, and that one year gave him an edge whenever their father pitted the two boys against each other to see who would win the coveted role of father's favourite by turning in the better performance at the piano, in mathematics challenges or in sports. More often than not, little Werner's defeat ended with him crying in the comfort of his mother's arms.[7]

Heisenberg grew up in a typically patriarchal German family. His father, August, was an ambitious and very busy man; his mother, Annie, a gentle, quiet woman who accepted her role as obedient wife and self-sacrificing mother. August fully intended to rise from high school teacher to professor, and the only way to do this was to publish a great many papers and become recognized in his field of middle and modern Greek studies. The family's lifestyle and activities were geared to advancing August's ambitions. The Heisenbergs adopted the appropriate religion, the most suitable philosophy and the kind of habits that would ensure August and his family had the appearance of "genteel respectability, social grace and allegiance to nationalist trappings."[8] Under August's command, they would do what was necessary to get ahead.

It was clear that August ruled the roost with a firm, authoritarian hand, but Annie was more than a quiet helpmeet running an orderly household. She was a smart and well-educated woman who graded the homework of her husband's students. She even learned Russian so she could translate academic papers for him.[9]

August finally acquired the coveted position of professor at the University of Munich when Werner was nine years old. Not much changed for Werner with the move. His father still worked long hours and was prone to stormy emotional outbursts, and Werner and Erwin continued to compete for their father's attention. According to biographer David Cassidy, the daily battles between the boys eventually degenerated into physical violence. "As they grew older, they fought even more frequently and intensely. Finally, after one particularly bloody fight—in which they beat each other with wooden chairs—they called a truce and went their separate ways."[10]

As a teenager, Werner found a role for himself that suited his love of the outdoors and nurtured his wounded self-confidence. He joined the German youth movement, modelled after the Boy Scouts in England, and found himself in charge of a troop of younger boys. As leader, Werner could play a role that was new to him: the kind and trusted older brother.

Heisenberg left high school and entered the University of Munich in 1921 to study science. He was highly motivated by the need to please his father, to live up to his father's ambitions for him and to emulate his father's drive to succeed. In Pauli, Heisenberg found a surrogate older brother, someone who could give him the kind of guidance and advice he needed—but who wouldn't beat him up. Once he was officially taken under Sommerfeld's wing as a doctoral student, Heisenberg blossomed. He was a handsome young man with a penchant for the "purity" of outdoor activities and sports, a popular theme in German culture. And he was brimming with new-found confidence.

When Sommerfeld decided it would be good for Heisenberg to attend the "Bohr Fest" in Göttingen with him, Werner was delighted. He was even more delighted to be able to write home to his family that not only had he met the famous Bohr, he and Sommerfeld had spent the whole morning with the great man at his luxurious lodgings. And then Bohr invited him to Copenhagen! Heisenberg was disappointed that he couldn't go, but for the moment it was enough to have been asked.

Sommerfeld was preparing to go to the United States for several months to teach, and while in Göttingen he arranged for some of his advanced students to study there in his absence. Max Born agreed to take four students, including Heisenberg, whom he was already considering as Pauli's successor as his assistant. Sommerfeld was quite insistent, however, that Heisenberg return to Munich to finish his degree. It was a heady time for the young man; each of the leading schools in the quantized atom was making a bid for him.

Heisenberg arrived in Göttingen in the fall of 1922, but his first stint at the university was not particularly pleasurable. With so many students flocking to Göttingen to study mathematics and science, Born was very busy. He already had nine doctoral students of his own, and now four of Sommerfeld's students. Heisenberg was lonely and depressed. It didn't make life more tolerable that the out-of-control inflation was making it difficult to get enough to eat, particularly for university students. Fortunately, someone (perhaps Born) had arranged for a subsidized lunch for physics students at a private home across from the institute. So they had at least one good meal a day, even if it rarely included meat.[11] Back in Munich, the Heisenberg family was having a hard time making ends meet, so there wasn't a lot of spare money to fund Werner. Born came to the rescue by offering him a paying job as his private assistant until Heisenberg had to return to Munich, even though he was still a student.

In the grim and hungry winter of 1922, Göttingen theorists, under the direction of Born, were trying to see if anything could be learned about the orbital model of the atom from studying the orbits of celestial bodies. Initially, Born had been less than enthusiastic about Bohr's orbital atom, but now he saw a way of advancing Göttingen's study of the atom through celestial mechanics. Perhaps the idea of the atom as a planetary system might be more than just a useful metaphor. "The thought that the laws of the macrocosmos in the small reflect the terrestrial world," said Born, "obviously exercises a great magic on mankind's mind, indeed its form is rooted in the superstition (which is as old as the history of thought) that the destiny of men could be read from the stars."[12]

Born rather fancied the idea that there was a certain cosmic symmetry between the orbiting bodies of the heavens and the orbiting bodies of the atom. Every Monday evening, about eight of Born's assistants and advanced students would meet at the Born apartment to study Henri Poincaré's celestial mechanics. At first, Heisenberg didn't see the point of it all, but he soon got pulled into the discussions.[13]

Heisenberg was an excellent pianist, his talent honed in competition with his brother. Max had two Steinway grand pianos back to back in the drawing room, and there Werner and Max would launch into Schubert's music for two pianos or perhaps pieces that Max had specially arranged. Heisenberg soon became a regular guest at the Born home, and on some visits little Gustav Born could be found lying under the pianos, listening in wonderment as the music rolled over him.[14]

As the clock ticked down on Heisenberg's temporary stay at Göttingen, Born decided that keeping the talented young man was important for building the reputation of his institute. Heisenberg, Born told Einstein, "is easily as gifted at Pauli, but has a more pleasing personality. He also plays piano very well."[15] Furthermore, Born never had to send the maid to Heisenberg's lodgings to rouse him. So, early in 1923, Born sent a long letter to Sommerfeld, who was still in the United States, arguing persuasively that he needed Heisenberg more than Sommerfeld did and that Born would see to Heisenberg's habilitation at Göttingen after his graduation. Sommerfeld agreed.

Born and Heisenberg worked on the celestial model of the atom all through the winter. Born worked at home, and Heisenberg was there most days for music and meals. While the two men laboured at Born's desk over the mathematics of the celestial model, little Gustav collected the innumerable discarded sheets of paper filled with handwritten equations, and used the blank sides for drawing or made paper cut-outs and airplanes.[16]

By the spring of 1923, they were ready to give their model a test drive by seeing if it could predict the spectral lines for a helium atom. The model didn't work at all. It was a disaster. Deeply disappointed, Born and Heisenberg agreed that it was time to try some entirely new hypothesis or to rebuild the concepts of physics from the ground up.[17]

Born set to work doing just that. He was attempting to build the rules for quantum mechanics, as he started calling it, based on its own logic, rather than trying to crowbar it into the logic of classical physics. The mathematics of calculus was fine for continuous change in the classical world, but it didn't suit the discontinuous jumps of electrons. For that, Born used a different kind of mathematics (the calculus of differences) that built the "chunking" of energy right into the fundamental principles of quantum mechanics.

Heisenberg went back to Munich for his doctoral defence in July. Even with the failure of the planetary model, he was still on a high from his time spent in the rarefied realms of theoretical physics at Göttingen. He approached his defence with great confidence, but perhaps he should have kept in mind his father's warning in his early student days not to neglect experimental work. Heisenberg had done just that, and he completely blew the experimental part of the defence. One of his examiners decided that this "uninformed upstart, even if he were another of Sommerfeld's parade of prima donnas," was not deserving of a degree. After some heated arguing amongst the examiners, Heisenberg was awarded a degree, but with the university's second-lowest passing grade.[18] Mortified, Heisenberg went straight to Göttingen, hat in hand, to see if Born still wanted to take him on as his assistant in the fall after such a terrible performance. Born did. He wasn't about to let Heisenberg go so easily.

August Heisenberg, anxious that his son should succeed, suggested that maybe Werner wasn't really cut out for physics and would fare poorly in the tough competition for the few places at universities for theoretical physicists. Perhaps he was better suited for another line of work. But Werner decided that if Born had confidence in him, then he would trust Born's judgment.

As 1923 drew to a close, Born and Heisenberg continued working on the new quantum mechanics, although Heisenberg was advancing a variation of his own that used half quanta instead of whole-number quanta. Born urged him to publish, but August Heisenberg advised him to wait, demonstrating yet again his doubt about his son's abilities. Werner had not yet habilitated. If his model was wrong, it would only demonstrate his ignorance of physics. Heisenberg sent a copy of his manuscript to Bohr and Pauli in Copenhagen, downplaying his model as a "sausage with quantum sauce." Bohr seemed cautiously accepting of Heisenberg's model, but Pauli was not. He thought the model "ugly" and based on outrageous assumptions, and following his return to Hamburg, Pauli finally became so fed up with debating the details of the model with Heisenberg that he told Bohr he was not even going to bother working on quantum physics for the time being.[19] Heisenberg's article on the half quantum did not get published.

In the spring of 1924, Heisenberg made his first trip to Copenhagen during his Easter vacation. He ran into a little trouble with customs officials. "I knew no Danish and could not account for myself properly," he later wrote. "However, as soon as it became clear that I was about to work in Professor Bohr's institute, all difficulties were swept out of the way and all doors were opened for me."[20]

Unfortunately, Bohr and Kramers were in the thick of the BKS paper when Heisenberg arrived. Furthermore, Bohr was tied up with plans for the new house and laboratory next to the institute. It didn't look as if Heisenberg was going to see much of Bohr—but Bohr had other plans. He invited Heisenberg on a hundred-mile hike—just the two of them—in the countryside around Copenhagen.[21] This was perfect for Heisenberg because, not only did he love hiking in the countryside, the hike ensured he had Bohr's undivided attention for the duration. And since Bohr was fluent in German, language wasn't an issue.

Like Pauli, Heisenberg found Bohr's intuitive approach to quantum physics more appealing than Born's strong adherence to mathematical formalism. And just maybe the fact that Bohr had a Nobel Prize—and Born didn't—weighed in favour of spending more time in Copenhagen for a young scientist with big ambitions. Striding through the Danish countryside, Bohr and Heisenberg hatched a plan for the younger scientist to return in the fall.

While Heisenberg was away, Born was once again knocked out of commission by an asthma attack that left him bedridden for nearly a month. He returned to the university just as Copenhagen's new BKS theory was announced. It seemed that the Copenhagen group had temporarily taken the lead by offering a new theory for the quantized atom, but in truth they were all scrambling to make some kind of sense of the quantum. More worrisome to Born was the fact that Bohr wanted Heisenberg to come to Copenhagen for the winter. Born did not want to lose his assistant, but since he was organizing a lecture tour of America for the winter, it would have been churlish to refuse. He graciously ceded Heisenberg to Bohr, but only for seven months.

A few months after the Copenhagen physicists published the BKS theory, Born published "On Quantum Mechanics," which sketched out the new Göttingen mathematical approach, including the discrete, non-continuous mathematics that incorporated electron transitions between orbits. Thus, in the summer of 1924, the Copenhagen and Göttingen schools each had a new theory on offer.

Heisenberg headed back to Copenhagen in the fall, having

completed his habilitation in July with Born. He'd arranged to arrive when he knew Kramers would be away and he could fill in as Bohr's assistant. Kramers may not have been pleased to have Heisenberg occupying his desk, even if it was only for two weeks. Heisenberg was nipping at his heels, competitively speaking, and Kramers responded by treating his younger colleague with condescension and disdain.

A showdown between the two seemed inevitable. Kramers was working on a paper that used the usual orbital analogies to explain the emission and absorption of light, essentially a follow-up to the BKS paper. However, Heisenberg thought Kramers should apply Göttingen's new mathematical formalism and skip the analogies, *and* put Heisenberg's name on the paper. It was a clash of personalities, but also a clash between the visualizability of the orbital model of the atom, championed by Kramers, and the non-visualizability of mathematical symbolism, put forward by Heisenberg.

As Christmas neared, the two finally took their dispute to Bohr. He had to decide which approach was the best and whose name should go on the paper. Bohr, after listening to each man argue his case, ruled that Heisenberg's mathematical approach would be used and that his name would go on the paper as co-author.[22] It seemed "Father" Bohr had a new favourite.

Born ended up postponing the American trip he'd envisioned for the winter of 1924. He wasn't planning to replace Heisenberg with a new assistant, but he did have his eye on a promising young mathematician who had arrived at Göttingen the year before, as so many German science and mathematics students were wont to do as part of their university education. Pascual Jordan was a year younger than Heisenberg, but he had already begun establishing his credentials as a new kid on the block within the powerful mathematics dynasty at Göttingen.

Not long after he settled there, Jordan was awarded the coveted position of assistant to mathematician Richard Courant. Courant had got his start under the direction of the famed Hilbert and Klein, just as Born had. Courant had also served as assistant to Hilbert, only a few years after Born had held the same position. Born and Courant were thus well established within the Göttingen hierarchy as second-generation mathematicians of the Klein–Hilbert dynasty. Jordan was set to become one of the third generation.

Courant was busy writing a book on mathematical methods in physics and facing a looming deadline when Jordan arrived. It was the student's job to write up Courant's lecture notes, the same notes he was using for his book. Courant and Hilbert were listed as the book's authors, but Hilbert was too busy to do more than pass on his own lecture notes to Courant and Jordan.[23] It was an excellent exercise for Jordan, particularly since it gave him the opportunity to become quite familiar with the first chapter's abstract mathematics—linear algebra. This was a complicated method of multiplying a matrix of, say, figures in four rows and four columns by another matrix of four rows and columns. It was certainly not a well-known form of mathematics, even though it had been around for nearly fifty years.

Jordan, only twenty-one in 1924, was also immersed in the writing of another book that blended physics and mathematics, this time as co-author with the recently arrived head of Göttingen's experimental physics, James Franck, Born's counterpart at the physics institute. At the same time, Jordan was working on his doctoral degree under Born's supervision. For this, he had taken on the challenging task of bridging the gap between Einstein's light quanta and Copenhagen's new BKS paper, which, of course, was written for the express purpose of doing away with light quanta.[24] All in all, it was a challenging start for the Bavarian student.

Jordan was a shy young man, owlish behind his wire-rimmed spectacles. His pronounced stutter made it difficult for him to hold a conversation, but what he lacked in verbal skills he more than made up for in the mathematics department. By all outward appearances he was mild-mannered and slight, but beneath that demeanour beat the passionate heart of a devout nationalist and religious fundamentalist.

If he wanted to, Jordan could boast a noble ancestry. His great-grandfather, Pascual Jorda, was descended from an ancient line of Spanish nobility. He'd ended up in Hanover after the British and Spanish defeated Napoleon under the leadership of Wellington, and he stayed on to work for the British government. The family eventually changed their name to Jordan, but they retained the tradition of giving the first-born son the Spanish name of Pascual.[25]

Unlike his siblings, who were Catholics, Pascual was brought up in the Lutheran faith. He developed an early and deep interest in Protestant religion, undergoing a period of intense soul-searching when he was twelve, as he tried to reconcile the Bible and church teachings with Darwinism and modern science. Relief came with the pronouncement by his teacher of religion that there was really no contradiction between religion and science.[26] Jordan accepted this idea wholeheartedly. He could have his science and his religion too.

The defeat of Germany in the war—a defeat that many Germans couldn't accept because no battle had been fought or lost on German soil—fuelled Jordan's nationalist sentiments. Right-wing nationalism had never disappeared from the German scene, even with the demise of the Prussian military elite, but food shortages, inflation and the shakiness of the government of the Weimar Republic had given rise to a much more extreme and brutal form of nationalism.

For all his nationalist passion, however, Jordan did not appear to embrace the anti-Semitism that coloured so much of the right-wing nationalist thinking. If he had, he would have been much more likely to take his doctoral degree at the university in Munich, which had become home to a particularly vicious form of hatred towards Jews. Instead, he had chosen to work closely in Göttingen with Courant and Born, who were both Jewish.

Jordan's nationalist passion seemed to be fed by anti-Communism. The October Revolution, which swept the Russian monarchy from power to replace it with a Communist government headed by Lenin, had deeply alarmed him, and his attraction to Germany's National Socialist Party appeared to stem from the party's rabid anti-Communism rather than its anti-Semitism.[27] Jordan kept his strongly held views to himself, at least around his scientific colleagues. Perhaps he did not want to antagonize his academic superiors at the vulnerable beginning stages of his career.

In the summer of 1924, as Jordan beavered away at his doctoral degree and notes for the book on physics and mathematics, another reserved young mathematician was quietly working on his degree in Cambridge.

Paul Dirac, only a month or so older than Jordan, had come to Cambridge on a scholarship in 1923, fully intending to work on Einstein's general theory of relativity, then considered in England to be a form of applied mathematics. Dirac was quite disappointed when his preferred supervisor decided he already had enough students to look after and passed him along to Ralph Fowler, Ernest Rutherford's son-in-law. Fowler, not surprisingly, was far more interested in the theory of the quantized atom than in relativity.

Dirac had a very bright mind, hidden behind a thick shell of reserve, and he was a quick study. His earlier training at the University of Bristol had left him ill prepared for the advanced level of research at Cambridge. He knew very little about atomic theory, and the latest developments came as quite a surprise. As Dirac recalled:

> Fowler introduced me to quite a new field of interest, namely the atom of Rutherford, Bohr and Sommerfeld. Previously I had heard nothing about Bohr theory, and it was quite an eye-opener for me . . . The atoms were always considered as very hypothetical things by me, and here were people actually dealing with equations concerning the structure of the atom.[28]

Fortunately, he knew enough German that he could work his way through the research papers coming out of Munich and Göttingen. Within six months of arriving at Cambridge he produced two papers on statistical mechanics, and in May 1924 he wrote his first paper on problems in quantum mechanics.

Although Dirac and Jordan were the same age, Dirac had taken a somewhat more circuitous route to arrive at atomic physics. Dirac's father grew up in a French-speaking canton in Switzerland, but he cut off all contact with his parents when he was twenty and ended up in Bristol, England, making a living as a French teacher. Charles Dirac was a linguist with a remarkable talent for languages, but he was also reputed to be a hard man and a domestic tyrant with little understanding of how to build a relationship with his sons. Paul was the middle child; his brother Reginald was two years older, his sister Beatrice four years younger.

This was not a chatty family. In an effort to ensure his children learned to speak the language of their Dirac ancestors, Charles decreed they must speak to him in French as much as possible. He also insisted that only grammatically correct French was acceptable at the dinner table, and if the children didn't comply, they were punished. Paul was more fluent in French than his siblings, so he was allowed to eat with Charles in the dining room, while Reginald and Beatrice ate in the kitchen with their mother. Since Paul feared making a grammatical slip and incurring his father's wrath, he learned to keep quiet and not say anything at all.[29]

However many languages Charles could speak (possibly eight or nine), he couldn't seem to communicate with Reginald and Paul. The two boys didn't talk either. When Paul was eighteen, he took a summer job at the same factory where Reginald worked, but if the boys' paths happened to cross, they did not exchange a single word.[30] Charles insisted his elder son become an engineer, although Reginald really wanted to be a doctor. Paul didn't have any particular ambition and simply followed the same path as his brother.

Paul had been accelerated through school, so he started college when he was only sixteen. It wasn't, he said, that he was an unusually gifted student; it was simply that the college classrooms were emptying as young men went off to fight in the war, so the younger boys were pushed through quickly so that the professors would have somebody to teach.[31]

Paul had long taken refuge in math books. Mathematics was something he was very good at, but not something to be considered as a possible career. Nonetheless, after finishing his engineering degree in June 1921, Dirac took a scholarship examination to study mathematics at Cambridge, and he was successful. But his father could not afford to cover the rest of his expenses, so that plan was scrapped. Fortunately, Dirac was offered free tuition by the staff at the University of Bristol mathematics department, who were quite happy to take on a student they considered to be particularly talented. There he was introduced to the world of pure mathematics, the kind that was unsullied by any attempt to apply it to physics or any other practical problems of the world. The physics department at the university was in a different building, but Dirac made a point of attending some lectures there, and he became intrigued by general relativity. He did well at Bristol and was rewarded with enough scholarship and grants funding to finally study at Cambridge.

At the age of twenty-one, Dirac was finally free of his father's influence, but he didn't suddenly come out of his shell. Dirac was in the theoretical physics group at Cambridge, which didn't have its own building. Students ended up working in their rooms or in a small library/tea room at Cavendish Lab, so his isolation continued. He did, however, attend a weekly tea at the home of one of the professors, always dressed in a dark suit, and joined two physics clubs.[32] He also made a point of taking long, solitary walks in the country on Sundays. He loved mountain walking as well, and could sometimes be seen climbing trees in the hills near Cambridge as part of his training, dressed as always in his dark suit.[33] Otherwise, he avoided anything that might distract him from his studies, such as politics, sports and girls.[34]

On March 5, 1925, Reginald killed himself in a field in Shropshire, where he was working as a draftsman, perhaps because he felt stuck in a career he didn't care for, or perhaps because he was struggling with relationship problems with his girlfriend.[35] Charles Dirac took his son's death badly and was so emotionally distraught that Paul feared his father might go quite mad. Paul, vowing never to consider such a drastic and destructive end for himself no matter what the circumstances,[36] retreated a little more into his shell and continued at Cambridge, producing respectable if unremarkable papers.

Another doctoral student in theoretical physics with a string of unremarkable papers behind him was Prince Louis de Broglie. Pauli, Heisenberg, Jordan and Dirac were all born within a few years of each other, and they had all been young enough to avoid being called up for active service in the war. Not so for de Broglie, who was twenty-one when the war started. He'd finished his undergraduate science degree at the Sorbonne and had just signed up for his mandatory military service in the French army when the war broke out. De Broglie's older brother Maurice, the sixth duc de Broglie,[37] pulled a few strings and got Louis stationed in Paris, where he spent the entire war as a telegraph operator at the base of the Eiffel Tower.

Immediately after leaving the army in 1919, he headed back to the Sorbonne, attending a seminar by Paul Langevin on quantum theory and working with his brother in his well-equipped laboratory in his Paris townhouse. As he later recalled: "Demobilized in 1919, I returned to the studies I had given up, while following closely the work pursued by my brother in his private laboratory with his young collaborators on X-ray spectra and the photoelectric effect. Thus I made my first steps in research by publishing a few results in the fields studied by my brother."[38]

Louis benefited greatly by working with Maurice, a well-respected experimentalist, but he became more and more intrigued by the theory behind the puzzle of the quantum, and that became the focus of his doctoral thesis at the Sorbonne.

Louis hadn't always been interested in physics. As a child he had seemed ideally suited to carry on the family's distinguished history of producing some of France's leading politicians and diplomats. The family's title had been bestowed in 1740 by Louis XIV, along with a castle and a vast region in Normandy that took the name Broglie. Most (but not all) of the family survived the French Revolution. When his father died in 1909, Maurice became the sixth duc de Broglie and young Louis's guardian.

Louis's early years were, quite literally, those of a fairy tale prince living in a castle. He was the youngest of five children, and much younger than Maurice, but he had his sister Pauline, who was only four years older, as his indulgent childhood companion. According to Pauline, little Louis just couldn't shut up. Having been raised in relative loneliness and schooled at home by private tutors, once Louis had an audience, he had to perform. As he grew up, he expanded his repertoire. As Pauline recollected:

> He had a prodigious memory, and knew by heart entire scenes from the classical theatre that he recited with inexhaustible verve . . . Hearing our parents discuss politics, he improvised speeches inspired by the accounts in the newspapers and could recite unerringly complete lists of ministers of the Third Republic, which changed so often . . . a great future as a statesman was predicted for Louis.[39]

But the irrepressible Louis ran into difficulties in university, first getting a degree in history, then looking to philosophy and law, and finally settling on physics. Unsure about a career direction, he failed a physics exam and went into a deep depression. According to Maurice, "Nothing looks promising, is he to become a failure in spite of his labours? Gone the gaiety and high spirits of his adolescence! The brilliant chatter of his childhood has been muted by the depth of his reflections."[40]

Perhaps to cheer up his disheartened brother, Maurice took Louis with him to the First Solvay Conference in Brussels in 1911, where Maurice was serving as a scientific secretary. It's not clear if Louis attended any of the sessions, since he was not an invitee, but he may have rubbed shoulders with Einstein, Lorentz, Curie, Langevin and Rutherford at breakfast or dinner. Louis pored over Maurice's notes on the discussions at the conference on the quantum, and that seemed to snap him out of his funk. Inspired, he decided that he wanted to be a theoretical physicist, and completed another undergraduate degree in science in 1913.

Louis returned to university at the age of twenty-seven after his war service ended in 1919. The university physics courses at the Sorbonne, unfortunately, did not cover the more recent advances in physics, such as Bohr's theory of the quantized atom, and many textbooks that were readily available for German physics students were not available in France. But de Broglie was fortunate: not only did he have an older brother working on cutting-edge experimental physics, he could take in all of Henri Poincaré's lectures on electrodynamics, thermodynamics and celestial mechanics.

De Broglie may, however, have developed a somewhat overblown sense of his abilities as a theorist. One of his early faux pas was to reinterpret Niels Bohr's pride and joy, the correspondence principle, and publish it in a French physics journal in 1921. Ever protective of Bohr's interests, an irate Kramers delivered a none too subtle slap-down in the English journal *Philosophical Magazine*, the same journal in which Bohr made a point of publishing the English translations of his papers.[41] It was just the beginning of The Copenhagen School versus The Annoying French Upstart, since de Broglie seemed deliberately to be taking a stance in opposition to Copenhagen thinking in general and to Bohr in particular. Even the theories of the Munich school led by Sommerfeld were a target for de Broglie's condescending dismissal, and it didn't help de Broglie's reputation in physics circles that sometimes the theories he published were simply wrong.

One of the papers de Broglie wrote in 1922 focused on the puzzle of wave/particle duality. "Then a great light dawned on me," he recalled later. What if he generalized Einstein's theory of the light quanta so that the motion of any particle—the light quanta or electron or any other particle—had a wave associated with it as well? Where Einstein had assigned the properties of particles (light quanta) to the light wave (radiation) back in 1905, de Broglie now assigned the properties of radiation to particles.[42]

De Broglie published three more short papers on the same subject in a French journal in the fall of 1923, and then extended and refined the idea for his doctoral thesis. In the spring of 1924 he took a draft of his thesis to Langevin to see if it needed any changes before its official submission. Langevin may have had his doubts about de Broglie's take on wave/particle duality, given de Broglie's somewhat sketchy track record. "M. Langevin, (probably a bit astonished by the novelty of my idea)," said de Broglie, "asked me to furnish him with a second typed copy of my Thèse for transmittal to Einstein. It was then that Einstein declared, after having read my work, that my ideas interested him. This made Langevin decide to accept my work."[43]

What intrigued Einstein was that de Broglie had made a direct link between matter, waves, energy and the quantized atom. He had taken the properties of a particle (energy and momentum), married them to the properties of a wave (frequency), and plugged in Planck's constant *h*, *et voilà*, de Broglie had produced a simple formula to show that all matter could be considered as a wave. This was something Einstein could work with, and he did.

The young men in theoretical physics—the boys' club—were just beginning to show a little muscle by 1924, but veterans like Einstein were hardly out for the count.

Chapter Ten

THE GÖTTINGEN GOSPEL

*Mathematics is not a careful march down a well-cleared path, but
a journey into a strange wilderness, where explorers often get lost.*

W. S. ANGLIN, mathematician

Albert Einstein was no less frustrated than Niels Bohr or Max Born
with the continuing groping in the dark to solve the puzzle of the
quantum. He'd stopped publishing papers about light quanta in
1917. It wasn't that he'd stopped thinking about them, but there
didn't seem to be anything more he could add to the discussion that
hadn't already been said. Einstein, however, got an unexpected boost
in the summer of 1924, mainly because he was quite conscientious
about reading and answering his mail, even what other scientists
might have considered crank mail.[1]

In June, Satyendra Nath Bose, an Indian physicist from a university in Dacca who was unknown in Europe, sent Einstein a copy
of a paper he'd tried to get published in the British journal *Philosophical Magazine*. The editor had rejected it. Bose asked Einstein
if he would read it and, if Einstein considered it worthy of publication, translate it from English to German and submit it. This might
have seemed a somewhat presumptuous request from a stranger,
but, as Bose pointed out, he wasn't a complete stranger: he had
sought permission from Einstein to have his papers on general relativity translated into English and published in India. But that's not
what caught Einstein's attention.

In his paper, Bose was offering a new take on the light quanta. He had been able to derive Planck's distribution law—the formula from 1900 that accounted for the distribution of quantized radiation frequencies inside a container—by using only Einstein's theory of light quanta. That had never been done before. A few other scientists had worked out different means of arriving at Planck's distribution law, but they had all used classical radiation as a starting point, just as Planck had. In contrast, Bose's derivation used only Einstein's light quanta and statistical mechanics, which gratified Einstein no end. It was, finally, the long-awaited advance of Einstein's work of 1916, when he had confirmed, at least to himself, the reality of light quanta. A pleased Einstein wrote to tell Paul Ehrenfest that "the Indian Bose has given a beautiful derivation of Planck's law . . ."[2]

To arrive at the light quanta version of Planck's law, however, Bose had made two bold leaps in the dark. First, Bose suggested that light quanta had two states of polarization (a kind of two-handedness), which had never been observed in particles before. And second, he assumed light quanta were not distinguishable from each other. Unlike electrons, which had strict "rules" about who could sit where in an atom, Bose's light quanta were very chummy. They were all the same, so they could, in theory, all sit in the exact same spot (or state). This made a big difference in the formula for calculating their distribution, because obviously particles that could "sit" anywhere had a different probability of being in a certain allowed state than particles with a long list of rules about who could sit next to whom.

It's rather like several games of musical chairs going on at the same time. Electrons, dressed in T-shirts with colours and stripes indicating their different characteristics within the atom, could sit only in groupings of 2, 8 or 18, for instance, and no two electrons with identical T-shirts could sit in the same group. These rules made it a much tougher exercise for everyone to find a place in one of the groupings where they were allowed to sit. Light quanta, on the other hand, would all be dressed in identical T-shirts, and could sit anywhere they wanted within any grouping. They could all pile up on the same chair if they wanted to.

By introducing a new way of counting light quanta, Bose was finally able to shed some light on the behaviour of light particles, and show that they didn't behave the same way as other particles, such as electrons, atoms or protons (the positively charged particle in the nucleus of the atom that Ernest Rutherford had identified in 1920).

Einstein was quick to notice that, with Bose's procedure, one could now imagine not only a container filled with radiation (Planck) or gas molecules (Boltzmann), but also a container filled with a gas of light quanta.

Einstein promptly translated Bose's paper himself and sent it off for publication, along with a note to the editor giving it his endorsement. "In my opinion Bose's derivation of the Planck formula signifies an important advance. The method used also yields the quantum theory of an ideal gas, as I will work out in detail elsewhere."[3] Of course, Einstein's endorsement pretty much guaranteed publication.

Within a week of sending off Bose's paper, Einstein presented his new quantum theory of an ideal gas to the Prussian Academy. For this, Einstein imagined a container of identical molecules, and showed that Bose's calculations for indistinguishable light quanta also gave the correct results for indistinguishable molecules. The same applied to Planck's radiation if one considered only a single frequency. Here was a strong indicator that, in some way, radiation, material particles such as molecules, and light quanta were very similar.

Enter Louis de Broglie. His wave theory assumed that if a wave could be a particle, a particle could also be a wave. And not only did light exhibit properties of wave/particle duality, so did matter. It's not entirely clear when Einstein learned of de Broglie's theory.*

---

*There is some question as to when Einstein first received the draft copy of de Broglie's thesis, and whether it was sent to him by Langevin or de Broglie.

By the summer of 1924, de Broglie had published three papers on the subject in a French journal, but French journals were not readily available in Germany, and vice versa. De Broglie's thesis jury at the Sorbonne had something of a problem. Just as Niels Bohr's examiners in 1911 had found themselves unable to challenge the student defender because his work was too far out of their range of knowledge, so did de Broglie's. Paul Langevin, then a professor at the Collège de France, served as the invited examiner. He was the only one of the examiners who had a depth of knowledge about quantum physics, and even he wasn't sure about de Broglie's thesis. They all knew and respected the work of Maurice de Broglie; they just weren't so sure about his younger brother. Langevin, carrying a copy of de Broglie's thesis under his arm, told a colleague, "I am taking with me the little brother's thesis. Looks far-fetched to me."4 But Einstein could see the farther-reaching implications of de Broglie's thesis and gave Langevin a most complimentary review. De Broglie was duly awarded his doctoral degree in November 1924.

Einstein had given a talk at a physics conference that fall in which he proposed an experimental approach for testing for a gas of light quanta similar to the experimental suggestion de Broglie had made in one of his earlier papers, which Einstein was likely to have seen by then. As 1924 ended, Einstein was preparing a paper on the quantum theory of an ideal gas, which was published the following January. In a reflection of de Broglie's own conclusion, Einstein noted in the preface of the paper, "If one is justified in considering radiation as a gas of quanta, then the analogy between the gas of quanta and the gas of molecules must be complete."5

Einstein's results, using the work of Bose and de Broglie, seemed to confirm that the paradox of wave/particle duality for radiation was not just an annoying problem that physicists had not yet figured out how to solve; it was an inherent property of the quantum world. Even though the physics community tended to pay close attention to what the great Einstein had to say, scientists were hardly falling all over themselves to rush off and test this new hypothesis. However, Cambridge student Paul Dirac did send a letter to de Broglie requesting a copy of the thesis because he was planning to give a lecture on the subject.[6]

Scientists who were familiar with de Broglie's work, including those at Copenhagen and Munich, had reason to be very skeptical. One of Sommerfeld's doctoral students in Munich noted that "de Broglie's paper was studied there too. Everyone had objections (they were not very difficult to find) and no one took the idea seriously."[7] De Broglie had earned himself a reputation in scientific circles of being a cross between a crank and a reactionary,[8] and it took more than Einstein's endorsement to persuade his colleagues to have second thoughts. And besides, most theoretical physicists already had quite enough on their plates trying to make sense of wave/particle duality for radiation;[9] they weren't really in the mood to handle a curveball that forced them to look at wave/particle duality for matter as well.

But not everybody was ignoring de Broglie's matter waves. Max Born studied de Broglie's wave theory in the spring of 1925, at Einstein's urging, and one of James Franck's students was pondering it as well. Walter Elsasser, fresh from Sommerfeld's "nursery" in early 1925, had attended a seminar given by one of Born's many students. The student, at Born's suggestion, had been looking into the puzzling experimental results reported by two American physicists working for Bell Telephone Laboratories. They had been firing electrons at a metal plate and the results looked a lot like a wave interference pattern, which didn't make any sense to them. Elsasser didn't think much more about the idea until he came across Einstein's papers from a few months earlier on the quantum theory of gas. Einstein had cited de Broglie's thesis, and although Elsasser thought it highly unlikely that he'd find a French thesis in a German library, he went looking for it anyway, and found it. He supposed that de Broglie had sent it to Born, who, being very busy, had merely glanced at it before sending it on to the library.[10]

Elsasser quickly made the connection between the Bell Telephone experimental results and de Broglie matter waves and, using only a slide rule and de Broglie's wave equation, worked out a rough calculation that matched the experimental results closely enough to make him wonder if he was on to something. Elsasser took the results to Franck, who thought the idea quite speculative but nonetheless encouraged the 20-year-old to write it up and submit it for publication.[11]

In the midst of it all, Born sent Einstein a long letter on July 15, 1925, bringing him up to date on what was happening in Göttingen. Born and Einstein had had a bit of a falling-out the previous year over the direction quantum physics was taking. Born had come to believe that the quantum world was inherently probabilistic and that quantum events, such as quantum jumps, were random and therefore without cause. Einstein disagreed. He wanted a physical theory that was based on an objective and visualizable external world, not one that was fundamentally probabilistic, acausal and random. At any rate, by that summer they'd mended fences again.

In his letter, Born first mentioned his difficulty in grasping Einstein's work on the Bose gas, but added that Ehrenfest had then showed up for his usual summer visit to Göttingen and was able to "cast some light on it. Then I read Louis de Broglie's paper, and gradually saw what they were up to. I now believe that the wave theory of matter could be of very great importance . . . I am also speculating a little about de Broglie waves."[12] He also mentioned that "our Mr. Elsasser" was still trying to get his thoughts in order regarding matter waves.[13]

Just three days after Born penned his long letter to Einstein,[14] Elsasser wrote up a short note on a way to test electrons to see if they were also waves. If electrons were waves, then when they were fired through a very fine grating they should exhibit the same kind of wave interference that happens when ripples radiating across a pond cross another set of ripples. When the high points of the ripples meet, they create higher crests, and when a high point of one wave meets the trough of another, they cancel each other out. This was a familiar experiment for testing the wave nature of light; Elsasser figured it could also demonstrate the wave nature of electrons.

Franck approved the short paper, and it was sent off for publication in a prominent German journal. In the paper, Elsasser argued that Einstein's work on the possibility of a wave field caused by the motion of particles was remarkable and warranted testing. "The hypothesis of such waves, already advanced by de Broglie before Einstein, is so strongly supported by Einstein's theory that it seems appropriate to look for experiments to test it."[15]

Einstein reviewed Elsasser's paper for the journal he'd submitted it to, and although Einstein wondered if one should really be taking the wave nature of electrons quite so literally, he recommended it for publication.[16] It was the first paper to appear in an academic journal explicitly linking de Broglie's matter waves with the experimental results.

Born did not have much time to speculate about de Broglie's waves or to follow up on Elsasser's experimental proposal, since he was in the midst of the summer visitor season. Ehrenfest had been to stay, and so had a number of Russian scientists. Hendrik Kramers had turned up, and so had Philipp Frank, the Prague physicist who was also one of the leaders of the logical positivism movement. Like some of the other Göttingen wives who sought refuge from the annual summer onslaught of visiting scientists and their entourages, Hedi had already taken herself and the children off on their summer holidays to Switzerland. Born had a whirlwind of activities ahead of him before he could join his family at the end of July. And he still had to take another look at the baffling paper Heisenberg had given him a week or so earlier, before Heisenberg left to lecture at Cambridge University, after which he would go on to Copenhagen to finish his scheduled time there.

It would have been a lot easier to figure out what Heisenberg was getting at in his paper if he'd shared what he was doing with Born, but he'd kept him in the dark. When the younger man had given him the paper, he'd been rather offhand about it, as if it wasn't anything particularly important. Heisenberg had taken it as far as he could, he said, and was prepared to leave it up to Born to decide if it was worth publishing.

Born was very tired, and Heisenberg's lack of enthusiasm didn't exactly inspire him to dig right into the paper. When he did, he found it difficult to read. Born certainly agreed with the basic premise of sticking to observables, but the math was puzzling, in part because Heisenberg gave so few clues about how he'd arrived at the calculations he was using.[17] And there were leaps in his reasoning that made it hard to follow his logic, although that was not surprising: Heisenberg was groping in the dark, taking a stab at a new approach, just as all the physicists working on the puzzle of the quantum had been doing for years. He didn't really know what he was doing. He hadn't worked out a firm foundation yet, but he was seeking to describe the behaviour of the quantized atom using only energy, intensity and frequency, and the observable experimental results from spectroscopy that showed up in the spectral signature of heated gases.

The study of spectral fingerprints had been going on since the spectroscope was invented in 1850, with scientists studying the radiation given off by the huge number of molecules in a heated gas through the spectroscope, which produced an array of coloured lines unique to each element. In 1925, spectroscopy was still the most commonly used means of testing atomic behaviour, but now physicists could make some statements with a certain degree of confidence: the coloured lines of the "fingerprint" represented the frequencies of emissions as electrons dropped from a higher energy orbit to a lower one; and the greater the intensity or brightness of the lines, the higher the number of electrons making the transition from one orbit to another at that frequency.

The brightest lines were a strong clue as to which orbits electrons were most likely to be in before they dropped to another orbit.

Einstein had already addressed this probability in his papers on light quanta in 1916 and 1917. But how to express the relationship between quantized energy and the transition probability—that was the puzzle Heisenberg was dealing with in his paper, assuming conservation of energy and using only observables. There was no longer any need to refer to invisible "orbits"; they were now allowed "states," and electron jumps from one orbit to another were simply "transitions" from one state to another. The unobservable orbits and quantum jumps were gone.

If this were classical physics, the power emitted during a transition would be proportional to the square of its position, and then the energy could readily be determined. But since an electron's position was unobservable—and hence could not be used in Heisenberg's calculations—he replaced it with the square of the probability of an electron making a transition from the initial state to the final one, although Heisenberg quickly recognized a kink in this correlation. An electron does not necessarily make a single transition from its initial state to the final state. It can transition from its initial state to an intermediate one and then to the final state, but just adding up the collective transition probabilities and then squaring the sum didn't work. The total energy of the quanta or radiation released must, by conservation of energy, be equal to the energy from the initial state minus the energy from the final state, and just squaring the sum of the probabilities didn't lead to the correct total energy.[18]

Heisenberg adjusted the mathematical rules to ensure that the probabilities were multiplied in the right order to get the correct solution for the energy, although this introduced a "significant difficulty" for Heisenberg.[19] In classical physics, it doesn't matter in which order one multiplies two quantities; the answer will always be the same. In Heisenberg's new quantum algebra, it did matter, because changing the order of multiplication produced two different answers. The quantities, in mathematical terms, did not commute with each other. It's rather like the order used when baking a cake—it matters if you add the eggs to the batter and then bake it, or if you bake the batter and then add the eggs.

And that's where Heisenberg left his summer paper, passing on copies to both Born and Pauli for their consideration. When Born had written to Einstein on July 15, he referred only briefly to Heisenberg's paper that would soon be published as appearing "rather mystifying but . . . certainly true and profound."[20]

What Born found mystifying were the rules for multiplication Heisenberg had used. They were vaguely familiar, but he couldn't quite place them. It really bothered Born that he, a mathematician by training, couldn't get past the niggling feeling that the mathematics in Heisenberg's paper were significant of something. He could barely sleep for thinking about it, which did nothing to help his general state of exhaustion. "Heisenberg's rule of multiplication left me no peace, and after a week of intensive thought and trial, I suddenly remembered an algebraic theory that I had learned from my teacher . . . in Breslau. Such quadratic arrays are quite familiar to mathematicians and are called matrices, in association with a definite rule of multiplication."[21]

Of course! It was the matrix multiplication used in linear algebra. It had been a long time since Born had used matrix multiplication, but he remembered it now. Back in 1908, Hermann Minkowski had used matrices to describe how radiation worked for moving bodies, and Born had used matrix multiplication the following year for a paper of his own on the motion of the electron.[22] If Born had shown Heisenberg's paper to Jordan, he would likely have recognized the matrix multiplication immediately, having worked with Courant on linear algebra only the year before.

Heisenberg had not studied the abstract algebra, but then, not very many physicists ever had. It was the domain of mathematicians, and most physicists were not mathematicians. Unwittingly, Heisenberg had plucked the non-commutation rule for multiplying matrices out of thin air, but because he didn't know that's what he had done, he could not put his observables-only model on the sound footing of the proven axioms of linear algebra. That was something Born could do.

On July 19, Born headed to Hanover for a meeting and ran into Wolfgang Pauli. Heisenberg had already sent Pauli a copy of the same paper he'd given Born, and Pauli had been thrilled. The new approach, he wrote to a friend, had pulled Pauli right out of the dumps and given him new hope. "Although it is not the solution to the riddle, I believe it is now once again possible to move forward."[23]

To Born, the mathematics in Heisenberg's paper obviously needed work, so he asked Pauli about collaborating with him on the changes. Pauli refused, responding in his typically blunt fashion, "I know you are fond of tedious and complicated formalism. You are only going to spoil Heisenberg's physical ideas by your futile mathematics."[24]

Back in Göttingen a few days later, Born went over Heisenberg's paper again and then sent it off for publication. But he knew the mathematics still needed sorting out. He forgot all about de Broglie waves and set to work putting a mathematical foundation under Heisenberg's paper, regardless of Pauli's scorn for such an effort. Indeed, Born was quite pleased at the prospect of using the advanced methods of matrix algebra. It fit neatly into his long-held belief that the problems in quantum physics would be resolved only through the use of a new mathematical tool. How convenient that this "new tool" was one he was already well equipped to use.

Born experienced a nagging doubt about whether he should be proceeding with the creation of matrix mechanics on his own while Heisenberg was in Copenhagen. He reassured himself that Heisenberg had handed him a completed paper, which would be published shortly, and thereafter every physicist would have the opportunity to develop his ideas. Born salved his conscience by deciding that the only difference was that he'd had access to Heisenberg's results a little before the others.[25]

Working quickly and independently, Born put together a paper that correctly identified Heisenberg's mathematical rule as stemming from matrix algebra, and clearly stated the rule that certain properties had to be multiplied in a specific order because changing the order of multiplication produced a different solution. The properties were therefore said to be non-commutating. Born was pleased that he was the first to write a physical law in non-commutating symbols,[26] but he couldn't work out a satisfying mathematical proof for it. For that, he needed Jordan's mathematical brilliance.

Born, although very weary, started working out the needed proof with Jordan, but Born had to leave to deliver a lecture on July 30, and then he was off to Switzerland to join Hedi and the children. Jordan, meanwhile, went to Hanover to spend some time with his parents. The two men continued their collaboration by mail, at least for a while. Jordan eagerly reported his progress to Born, but it turned out that Born had had a breakdown after he arrived in Switzerland. "After some time," Jordan recalled, "he considered it desirable to interrupt the exchange of correspondence since the double strain of a fatiguing sanatorium cure and our discussions through letters of this exciting subject did not agree well with him."[27]

Born didn't return to Göttingen until the middle of September, which left the lion's share of the work on developing the new matrix mechanics to Jordan, working alone at his parents' house. He had the notes Born and he had taken from Heisenberg's paper before it went off for publication, and he had the Courant–Hilbert book to help with the mathematics he needed. Jordan also wrote to Heisenberg, first at his parents' home in Munich and then in Copenhagen in September, to keep him abreast of what he and Born were doing.[28] The Born–Jordan paper went off for publication at the end of September.

Bohr did not know what was going on because Heisenberg, having learned a little caution from Pauli's often harsh critiques of his earlier ideas (and perhaps with the image in mind of his doubting father), did not tell him anything about his new approach until he arrived in Copenhagen in the middle of September. Bohr, who had been busy preparing lectures on the old quantum mechanics, which he considered to be at a "most provisional and unsatisfactory stage,"[29] was delighted when he got the gist of Heisenberg's idea. In a letter to Paul Ehrenfest, he declared, "I am full of enthusiasm [about Heisenberg's work] and generally about the prospects for further development."[30]

Pauli, however, was less than enthusiastic; he was downright snarky. He was quite dismissive of the mathematical formalism of the Born–Jordan paper. Yes, Pauli and Heisenberg had often mocked Born for his mathematical approach to physics, but this time Born's mathematics had produced the necessary foundation for Heisenberg's ideas. In doing so, Born had also handed Heisenberg his first big break in physics. Heisenberg now found the mathematics a lot more acceptable than he used to, and he finally blew up at Pauli:

> Your endless griping about Copenhagen and Göttingen is an utter disgrace. Surely you will allow that we are not deliberately trying to ruin physics. If you're complaining that we're such big jackasses because we haven't come up with anything physically new, maybe you're right. But then you're just as much of a jackass, since you haven't achieved anything either.[31]

That must have stung, given that Pauli was acutely aware of his perceived failure to come up with anything significant recently, and Heisenberg was just as aware of Pauli's vulnerability to dark self-doubt whenever he went through a fallow period. But the rebuke did spur Pauli on to tackle the hydrogen spectrum to see if the complicated matrix mathematics in the Born–Jordan paper might spit out the correct quantized energies.

It didn't take Heisenberg long to get up to speed on the complexities of matrix algebra with Jordan helping him. The trio of

Born and Jordan in Göttingen and Heisenberg in Copenhagen were putting together a mathematically logical and consistent quantum mechanics, which was exactly what Born had been aiming for all along. They were working under a tight deadline, however: Born was boarding a ship to the United States on October 26. Heisenberg returned to Göttingen at mid-month to help with finishing the paper and to get the necessary briefing from Born on helping run the institute in his absence. Born sent the paper off to the *Zeitschrift für Physik*, one of Germany's leading physics journals, where he was an editor.

Before leaving Göttingen, Born wrote to Bohr about the new quantum mechanics model. "I am so glad Heisenberg's idea pleases you. I believe with complete certainty that it signifies great progress and that the form, which Jordan and I have given it, is in a certain sense somewhat final, so far as one can say that at all in physics." Born was aware that Pauli wasn't the only one who thought he was too mathematically oriented, and he readily admitted to Bohr that he didn't have the same intuitive abilities as Bohr and Heisenberg. He added, "That is why I am so pleased that Heisenberg's theory fits so beautifully with my fixed idea and through it, one can get an overall view, as it appears to me, of all the theoretical possibilities."[32]

For Born, this was Göttingen's triumph. With the Copenhagen BKS theory shot down, the Göttingen model of quantum mechanics was now the leading model for the quantized atom. Bohr, however, had a more Copenhagen-centric take on things. The new matrix mechanics, he wrote to a colleague, was "resting upon the ideas of Kramers and especially of Heisenberg, shaped into such a wonderful theory by Born."[33]

Born was a little hesitant about leaving Göttingen just at the "birth" of the new matrix mechanics, but on the other hand American universities were eager to learn about the latest research in quantum physics. At Princeton, the faculty and students from the departments of physics, astronomy and mathematics were holding joint sessions to study Born's new theory in preparation for his arrival. Born knew better than to expect the kind of public excitement and hoopla that Einstein generated when he travelled, but he was heading to the United States at a triumphal moment and was excited about sharing the new Göttingen gospel with a receptive audience.

As Born left the university to go home and pack, Jordan handed him another paper he'd just written, asking if Born would forward it for publication in the *Zeitschrift für Physik*. Born popped the paper into one of his suitcases to read later and promptly forgot about it.[34] He was now looking forward to two weeks at sea with his wife, without the steady clamour of importuning students, the demands of visiting scientists, and piles of administrative paperwork to attend to. The children had been left in the care of Hedi's father. With so much of Max's time, energy and health devoted to building Göttingen's reputation for theoretical physics—and doing it so successfully—he had neglected his wife, and now Max had six months ahead of him to do what he could to restore his relationship with Hedi.

As for Heisenberg, no matter how he'd arrived at it, he'd come up with something really good, something that could establish him as a leading scientist and make his career. His father would be so proud.

Chapter Eleven

A MEETING OF MINDS

An institution is the lengthened shadow of one man.

RALPH WALDO EMERSON

In early December 1925, it was fair to say that things were getting very lively in Leiden. The joint was jumping, quite literally, in the university's physics department.

Two of Paul Ehrenfest's students had suggested identifying the two-handedness of Wolfgang Pauli's electron as a kind of "spin." It seemed that electrons could "spin" either clockwise or counter-clockwise. Leiden might not boast the intellectual depth of Göttingen's dynasties or the bevy of brilliant young scientists who orbited around Bohr at Copenhagen, but the university did have its share of talented students and professors. Still, it was a bit of a coup for Leiden to be the first.

The new "spin" was a bit of a comeuppance for Pauli. He'd started out the year, after his usual Christmas holiday in Vienna with his parents, by stopping on his way back to Hamburg at a small university in southern Germany. There, he'd been approached by a young physicist who thought he'd come up with a good idea to explain the two-handedness of the electron: spin. Pauli ridiculed the idea and so deflated the young scientist that he never showed his paper to anyone else.

In the spring of 1925, Pauli had published his paper on the two-handedness and exclusivity of electrons, and Samuel Goudsmit, one of Ehrenfest's doctoral students and an assistant, was puzzling over the significance of the quantum numbers Pauli had assigned to an electron. Another doctoral student, George Uhlenbeck, had just returned to Leiden from a tutoring job in Italy to resume his studies. Between the two of them, Goudsmit and Uhlenbeck came up with the idea of the electron as a rotating charge—hence the spin—and then produced a formula that seemed to explain all the aspects of the hydrogen spectrum.

From an electron's point of view, it was the electric field surrounding it that was rotating rather then the electron itself, just as, to people on Earth, the sky appears to rotate around the planet. Scientists had long ago shown that where there is a rotating electric field, there is a magnetic one as well, so Goudsmit and Uhlenbeck could explain the magnetic property of the atom as well as the spin, even though they did not have a lot of confidence in their solution. It seemed more like inspired guesswork or "a kind of numerology,"[1] so they took it to Ehrenfest. "Well, that is a nice idea," Ehrenfest told his two students, "though it may be wrong. But you don't yet have a reputation, so you have nothing to lose."[2]

Goudsmit and Uhlenbeck sent their short note off for publication on October 17, 1925, not thinking they had done anything particularly outstanding. A short time later, Goudsmit received a letter from Werner Heisenberg, congratulating them on their "courageous note."[3] It seemed the two students had stumbled onto something truly significant.

Pauli had little choice but to back down on his dislike of spin, but he could console himself that he had helped set the new matrix mechanics on a firmer footing. Still smarting from Heisenberg's earlier criticism of him—it was, after all, Pauli's role to criticize Heisenberg, not the other way around—Pauli had taken the time-consuming and laborious mathematics of the Göttingen matrix mechanics and correctly calculated the Balmer series for the hydrogen atom. It was the first demonstration that the matrix model might actually be able to explain the puzzle of the atom. It was much like when Bohr first came up with his quantized atom: only when Bohr had demonstrated to his own satisfaction that his model could produce the Balmer series for hydrogen was he prepared to take it seriously. On the downside, the new matrix mechanics was so mathematically complex that the challenge of calculating the details of the two-electron helium atom seemed impossible. Still, the hydrogen atom was a good start.

Niels Bohr heard about Pauli's success in reproducing the Balmer series at the end of November, and wrote to Pauli that he didn't know if he should be congratulating Heisenberg or Pauli the most.[4] Bohr learned of the Leiden results just as he was leaving Copenhagen for the Lorentz fiftieth-anniversary celebrations, and when his train to Leiden stopped at Hamburg, Pauli turned up at the station to ask his opinion of the spin. Bohr could see a potential problem with linking the already-established magnetic property of an electron with its spin, so his response was a cautious and non-committal "interesting"—which in Copenhagen-speak could also mean "nonsense."

When Bohr arrived at Leiden, Einstein and Ehrenfest met him at the station, and both immediately asked what he thought of the spin.[5] When Bohr again raised his concerns, Ehrenfest assured him that Einstein had already determined it was a relativistic effect, that the electron "sees" itself at rest while the electric field spins around it. "Upon my question about the cause of the necessary mutual coupling between the spin axis and the orbital motion," said Bohr a short time later, "[Einstein] explained that this coupling was a consequence of the theory of relativity. This remark acted as a complete revelation to me, and I have never faltered since in my conviction that we at last were at the end of our sorrows."[6]

The timing of the Leiden spin couldn't have been better. Ehrenfest was playing host to the many leading European theorists who had come to town to celebrate the long and productive career of Hendrik Lorentz. The elder statesman of physics had been a somewhat precocious young man, becoming the chair of physics at the University of Leiden, a position created just for him, at the ripe old age of twenty-five. His work with light and moving bodies had been put to good use in many areas of physics, especially Einstein's relativity. Lorentz had also headed up the committee that worked out the hydraulics and physics of Holland's massive reclamation of land from the sea in a complex diking system. There would be many notables in town to honour Lorentz, and Ehrenfest hoped it would mean Leiden's latest achievement in physics would get lots of attention too. But first there were festivities to attend to.

The many guests travelled by a special train from Leiden to Haarlem for the celebration, and on board were such international notables as Marie Curie, Ernest Rutherford and Arthur Eddington, along with Einstein and Bohr. Queen Wilhelmina of the Netherlands, in measure of the esteem in which Lorentz was held in their country, presented Lorentz with a gold medal in his name. All in all, it was a grand celebration for a man whose career had spanned the dramatic shift from the certainty of Newtonian physics to the puzzling world of the quantum.

When it was time for the dignitaries and guests to head home, Bohr and Einstein returned to Leiden. To Ehrenfest's delight, they had agreed to take up residence in the two upstairs bedrooms in the Ehrenfest house and were staying on for a week. Ehrenfest had vowed to leave the two Nobel laureates to their discussions, because the whole point of the get-together was to provide an environment where they could better understand each other's point of view and maybe find some common ground. "You realize what a great experience it would be for me to hear the two of you discussing the quantum riddles, but you can rely on me to leave you by yourselves almost all the time,"[7] Ehrenfest had written to Einstein.

Despite his best intentions, Ehrenfest didn't quite keep everyone away from Bohr and Einstein as he'd promised. With the implications of the new spin to be discussed, Goudsmit and Uhlenbeck were invited to Ehrenfest's house to explain it to Bohr and Einstein. The young scientists may very well have been a bit overwhelmed to be put on the spot by two of the world's leading physicists. As Goudsmit admitted, "When Bohr and Einstein were talking together at the Ehrenfests', I did not understand a bit of it."[8] Nevertheless, Bohr and Einstein did get their time alone to talk, and there was much to chew over: the new spin, the Göttingen matrix mechanics, the wave/particle nature of light and matter, and the beginnings of a disturbing trend of utilizing non-visualizable abstract mathematics in place of visualizable physics.

By the end of 1925, each of the three most important research centres in the study of atomic physics—Göttingen, Copenhagen and Munich—had established its own unique identity.

The Munich school, headed by Arnold Sommerfeld, had been working on the quantized atom for more than a decade. When he'd started, the study of the quantum barely existed as a field of research, and only a handful of people, such as Einstein, Ehrenfest, Bohr and some experimentalists, were devoting any amount of time to the topic. Sommerfeld encouraged his students to integrate mathematics, physics and experimentation to tackle a wide range of physics topics. Although he had a particular interest in the atom, Sommerfeld left it up to his students to decide if they wanted to delve into large philosophical issues in science. Such issues did not overly concern him.

Just before the end of the war, Sommerfeld published a book on the new atomic physics, and once the war was over, the University of Munich was flooded with eager students. Sommerfeld's seminars on the quantized atom were a big attraction, and the enthusiasm he inspired did not go unnoticed in the larger physics community. "What I admire about you," Einstein told Sommerfeld in 1922, "is that you have generated from nothing such an enormous number of young talents. This is something quite unique. You must have a gift to ennoble and activate the minds of your audience."[9] When Sommerfeld had taken over Munich's long-abandoned chair of theoretical physics in 1906, he had vowed to create a "nursery of theoretical physics,"[10] and he'd done just that. From the Sommerfeld nursery had emerged Heisenberg, Pauli and a raft of other young men eager to make their mark in atomic physics at Göttingen or Copenhagen.

By the early 1920s, Sommerfeld had developed his institute into a well-functioning and fertile physics program. Bohr, on the other hand, was just starting his own institute. There was no mathematical or physics dynasty in Denmark to build upon. But Bohr did have Ernest Rutherford's enterprise in Manchester to use as a template for the appropriate mix of theory and experimentation. Right from the start, Bohr intended his institute in Copenhagen to attract international scientists, which was a quite practical idea. Denmark had almost no physics community to speak of, so the theorists and experimentalists he'd need would have to come from outside the country anyway.

Bringing scientists to Copenhagen didn't turn out to be much of a problem. Not only did Bohr already hold one of the top honours in physics—a Nobel Prize—he was also very good at finding money to fund research assistants and visiting scientists. Copenhagen quickly became a popular destination for young scientists wanting to work at the forefront of atomic physics.

Taking another cue from Rutherford's practice, Bohr aimed for an informal and collegial atmosphere at the institute. He wanted to encourage creativity and interaction between students, young researchers and senior scientists. He wanted them to talk to each other. That desire was a reflection, perhaps, of Bohr's own need to talk out his ideas, sometimes to the point of exhausting the person he was with, who might or might not ever get a word in edgewise.

The Copenhagen institute was also a place of fun and relaxation, almost homey. Margrethe would serve sandwiches to students who were visiting with Niels, and their five blond-haired boys would dart in and out of the rooms. Visiting scientists and young researchers alike routinely went off to the cinema, played endless games of table tennis or visited cafés and bars.[11] This was the collegial, we're-all-in-this-together team spirit that Bohr had been aiming for.

The institute worked almost exclusively on quantum and atomic physics, and Bohr insisted that his correspondence principle be a benchmark for the students and young scientists. While Bohr encouraged a collaborative environment for research, he did not take kindly to being contradicted, as Kramers and Slater found out when they attempted to champion light quanta. Although he encouraged his assistants to challenge him, it was a daunting task, since "he rarely lost an argument," according to one of his assistants, "both when he was right and [when] he was wrong."[12]

Unlike Sommerfeld, Bohr had very specific philosophical ideas that governed the research program at his institute. But inasmuch as Bohr approached physics from an intuitive and philosophical perspective, he did not adhere to any particular school of philosophy and resented anyone who tried to attach a philosophical label to him.[13]

Bohr was a powerful figure at the institute, not only because it was *his* institute but because he had gathered a following of disciples who championed their leader to the point of worship. Bohr's considerable difficulty in enunciating his thoughts—his "suffering" during the thought process—became the hardship and martyrdom of a saint. Most of the senior physicists learned to keep their distance from Bohr and just got on with their work, while maintaining a balance between the appropriate humble reverence for Bohr and their own independence.[14] Bohr sometimes joked about being a prophet or guru, but he did not appear to seek out or deliberately cultivate such adoration. He took his responsibilities very seriously, and he did struggle mightily in trying to unravel the puzzle of the quantum. His struggles were seen as the mark of a prophet; thus, the Copenhagen school evolved into a temple for Bohr.

Unlike Bohr, Max Born didn't have to start from scratch. When he moved to Göttingen to fill the newly created position of chair of theoretical physics in 1920, he had the same freedom to create a new program as both Sommerfeld and Bohr, but there was already a tradition of mathematics-based theoretical physics at Göttingen as advocated by Felix Klein and, to a lesser degree, David Hilbert. Born's institute placed theoretical physics at the university in the physics department, essentially formalizing the marriage of physics and mathematics.

Göttingen already attracted a great many students, and Born's theoretical institute (and James Franck's experimental institute) was icing on the cake for the university's reputation. But the weight of Göttingen's traditions was as much a burden as a blessing to Born. In Germany, a professor's prestige was measured by how many students he supervised and how many papers were published under his direction. While Bohr and Einstein were protected from the need to supervise doctoral students—Bohr didn't want to and Einstein didn't need to—Born was expected to take on many students, as commensurate with his status.

According to physicist and philosopher Philipp Frank, the pressure to turn out large numbers of graduates and papers led many professors to accept less than exceptional students, who were given research projects that could be broken up into easily managed bits.

> In this way, there arose what was known in Germany as the *Betrieb* (mill), where to all outward appearances no distinction was made between worthwhile ideas and trivialities . . . An agreeable feeling of activity surrounded both teacher and students. They became so engrossed in this activity and industry that the larger problem that the partial studies were supposed to elucidate was often forgotten. The production of dissertations and papers became an end in itself.[15]

Such pressures were a problem for Born because he was intent on carving out a name for Göttingen as a world centre for atomic physics, but he was burdened with the responsibility for doctoral students, not all of whom could be called budding geniuses.

Because of the demands on his time, Born tended to appear remote, distracted and unavailable to his students. He preferred to work at home, and a student who wanted to talk to him had to phone several days in advance and make an appointment to see him.

Born did, however, gather around him a few students and assistants whom he considered worthy of his time and attention, people such as Jordan and Heisenberg. In particular, he wanted people around him who understood the value of basing physics on the solid foundation of mathematics. It is not surprising, then, that the mathematical formalism of the new matrix mechanics would emerge from Göttingen rather than Copenhagen or Munich.

While Sommerfeld might warrant appreciation from the students he nurtured in his Munich "nursery" and Bohr was the focus of adoration at the Copenhagen temple, Born suffered from a remoteness (and neuroticism) that seemed to preclude building a base of admiring supporters at Göttingen. Even so, he was at the vanguard of mathematics as it formalized its role in the domain of quantum physics. To all intents and purposes, Born had established the physics branch of the Göttingen mathematical dynasty.

The three main schools of atomic physics—Sommerfeld's theoretical nursery, Bohr's philosophical temple and Born's mathematical physics dynasty—were not the only places where research was being done, of course. But they were certainly the prime movers in the development of quantum physics.

Despite Einstein's celebrity status and Nobel Prize, he did not become the leader of any particular school of science or philosophy. In part that was because what Einstein chose to work on was usually too difficult for students to follow. He was better off working alone or with a few respected colleagues. Early in his career, when he was a professor at Prague, he had an open-door policy for students and encouraged them to talk about their work. He didn't mind being interrupted, he said, because he could easily pick up thinking where he'd left off. When he moved to Berlin, Einstein had less contact with students because he did not have to give lectures or supervise doctoral degrees, but he still needed someone to talk to about his ideas. According to Philipp Frank, Einstein had to clarify his ideas by expressing them out loud, not unlike Bohr. "Thus [Einstein] often conversed with students about scientific problems or told them his new ideas. But Einstein did not care whether the listener really understood what was being explained or not; all that was necessary was that [the listener] should not appear too stupid or uninterested."[16]

Because Einstein was not part of the German academic "mill," he actually had trouble finding assistants to work with him.* Doctoral students were far too busy jumping through the necessary hoops in order to get their degrees to be able to spend time on anything that did not advance their career prospects; ironically, that included working with Einstein. As a result, Einstein usually hired assistants from outside the country who were not exclusively focused on climbing the academic ladder in Germany.[17]

Ehrenfest's position in Leiden was more like Born's in that he was expected to supervise a large number of doctoral students as well as lecture and do research. Even with the financial pressures he was under, Ehrenfest had time for his students and made a point of being as helpful and encouraging as possible. He was a popular teacher, but, as always, he focused not on what he did well but on his perceived failures; so it was a particular pleasure for him to be able to engage Einstein and Bohr in discussion of the electron spin, a discovery made by his own students at Leiden.

Ehrenfest, Bohr and Einstein were all members of the old school of physics. While mathematics was certainly useful in physics, old-school physicists were not expected to be mathematicians. When

---

*One of Einstein's first assistants in Berlin, according to Philipp Frank, was a young man who couldn't get a job anywhere else because of his facial deformity. He suffered from leontiasis, sometimes called "lion face" because of the excessive growth of the facial bones.

Einstein and the others were at university, applied mathematics was generally just another term for engineering, but gradually mathematics had become more and more important in the work of physicists. Boltzmann's probabilistic mathematics was built right into the theory of the atom. Einstein's general relativity was supported by the somewhat unwieldy mathematics of tensor calculus. Nevertheless, it was not, to Einstein, abstract mathematics; he could visualize the curved space that the equations represented. In such old-school physics, the mathematics was used to support a visualizable idea. But the new matrix mechanics was different: its mathematics was intentionally circumventing visualizability.

Born, Heisenberg and Jordan had formulated quantum mechanics so that visualizable elements, such as position and momentum, were embedded in the abstract mathematical symbols of non-commutative algebra. The description of what was happening inside an atom was reformulated in the language of symbols, so that it was no longer possible to visualize what was going on because the symbols did not actually have a physical meaning. The atom had essentially been turned into a black box that no one could see into. For Born, a mathematician, that was quite acceptable. As long as the mathematical symbols, arranged in suitable combinations, spat out solutions that correctly predicted experimental results, what did it matter that the mathematics said nothing about the physics of what was going on?

Einstein, who was not a particularly great mathematician, once remarked, "The people in Göttingen sometimes strike me, not as if they wanted to help one formulate something clearly, but instead as if they wanted only to show us physicists how much brighter they are than we."[18] He wasn't far wrong. The venerable Felix Klein, who had died in the summer of 1925, was reputed to have wondered if physics was really too difficult for physicists and should properly be left to mathematicians.

Mathematicians, especially those who pursued pure mathematics, considered their work to be transcendent, beyond the shabby toiling of those who applied mathematics to the mundane problems of the world. Yet many physicists looked upon mathematicians as mere toolmakers who supplied them with a tool box filled with mathematical procedures, from which they could pick and choose the appropriate implements they needed to develop their theories. Each considered their discipline superior to the other. This elbowing for dominance in physics between mathematicians and physicists trying to explain nature and physical reality had been going on for a very long time. Even Newton's use of mathematics had set off alarm bells with some philosophers, who were worried that Newtonian physics was seducing physicists, and that mathematics "had been introduced into physics by people who were better at calculating than at explanations."[19]

Attitudes had changed over the intervening 175 years in the debate about the validity of mathematics in physics. Mathematics had been steadily working its way into all corners of physics, so that physicists in 1925 were needing to become much more adept at calculation. The advent of matrix mechanics was explicitly giving primacy to symbolic mathematics over visualizable physics. The "strict goddess of Science" was being slowly but firmly edged out of her domain by the expansiveness of the even stricter god of Mathematics.

It was not lost on Bohr that the new matrix mechanics, presented in Göttingen's style of abstract mathematics, did not bear much resemblance to the philosophy and physical reasoning that it had originally been based on. But in the aftermath of the failed BKS proposal, Bohr had already decided that something was going to have to go, although apparently it wasn't going to be light quanta or conservation of energy.

While the shift to mathematical formalism in quantum physics pleased Born no end because it was just his cup of tea, it did not suit Bohr at all. Bohr had never been particularly interested in mathe-

matics, and he relied on assistants such as Kramers to do the necessary math in support of his images and analogies.

Earlier in 1925, Bohr was giving serious consideration to leaving behind the usual space-time description of classical physics as a
means of describing how an atom behaved, and had been racking
his brains to figure out what analogies he could use instead. Born,
Jordan and Heisenberg had now provided a conception of the atom
that did indeed leave behind classical space-time descriptions, but
there was no analogy or metaphor to help one imagine what might
be happening inside the atom. The mathematicians might be happy
with their symbols, but Bohr was not. That wasn't how he did
science.

Einstein had as little liking as Bohr for the substitution of mathematics for physics. His success in science was based on his ability
to imagine a physics puzzle and work it around in his mind until
it made sense to him, and then he went looking for something useful in the mathematics tool box to support his idea. Still, the new
matrix mechanics was the best proposal anyone had yet come up
with in the search for a rational, rules-based quantum mechanics.
Unfortunately, it was based on probabilities and randomness, and,
worse, it was non-visualizable. Neither man could be satisfied with
a science that was leaning more and more heavily on mathematics
at the expense of an understandable explanatory theory.

Bohr and Einstein may have shared significant concerns about
the direction quantum physics was taking, but their time in Leiden
did not result in them coming up with a new approach or an agreement on what should stay from classical physics and what should
go. On that point they differed greatly. Nonetheless, they did have
time to talk and to understand each other's ideas very well, even
though they did not find the common ground that Ehrenfest had
hoped for. On the positive side, both men had the benefit of time
away from the demands of the outside world in the enjoyable company of the Ehrenfest family. In a thank-you note to Ehrenfest just

before Christmas, Bohr said his visit had been "a wonderful experience" and that his talks with Einstein were "a greater pleasure and more constructive than I can say."[20]

Born had missed the Lorentz celebration because of his trip to America, and as Christmas came and went he'd begun wondering if it had been wise to leave Germany just as matrix mechanics was finding its feet. The trip wasn't going quite as planned. Hedi had become so ill from a particularly nasty bout of food poisoning that she'd ended up in hospital, and she didn't seem to be recovering very well. And now Bohr had written to him that he wanted Heisenberg as his assistant in Copenhagen. After ten years as Bohr's right-hand man, Kramers had taken a position at Utrecht University, and Heisenberg was first in line to take over his desk. Born didn't want to lose Heisenberg, but he was a quarter of the way around the world and not in a position to negotiate with him. He didn't have much choice. Born wrote to Bohr in December:

> To lose him is for me a very great loss, we have worked together so beautifully and I like him personally so much—but it is a difficult loss for our institute and university . . . In any case, I must hand off Heisenberg to you, since for him it is still splendid to be recognized so young and to be allowed to work with you.[21]

Born had also had a bit of a shock earlier in the winter when he'd opened his mail and discovered a copy of a journal article with the title "The Fundamental Equations of Quantum Mechanics." It was written by one P.A.M. Dirac at Cambridge, someone Born had never heard of, and it contained essentially the same non-commutative algebra that Born had worked out in September with Jordan. Worse, this Dirac person had published his version of the mathematics for matrix mechanics in an English journal fully nine

Albert Einstein and Wolfgang Pauli, 1926, at the Ehrenfest home.

ABOVE: Max Born tweaks Wolfgang Pauli's ear during an unusually playful moment in 1925. BELOW: Wolfgang Pauli, Werner Heisenberg and Enrico Fermi in Como, September 1927.

ABOVE: Paul Dirac, far left, with visiting Russian scientists in Leiden, 1929. BELOW: Wolfgang Pauli and Paul Ehrenfest share a laugh on a boat trip, 1929

ABOVE: 1927 Solvay Conference. BELOW: Pascual Jordan, far right, 26, at a 1929 physics conference in the USSR.

Wolfgang Pauli in 1924, at age 24, a year of frustration.

The 1922 "Bohrfest" in Göttingen, with Swedish physicist Carl Oseen, Niels Bohr, James Franck, Oskar Klein and Max Born (sitting).

Niels Bohr and Albert Einstein in December 1925, relaxed and contemplative, at the Ehrenfest home in Leiden.

ABOVE: Erwin Schrödinger, at about age 20 in Vienna, following his return from military service in 1917. LEFT: Prince Louis de Broglie as a young man. BELOW: Paul Ehrenfest, with son Paul Jr. sitting on Albert Einstein's knee, at the Ehrenfest home in Leiden in the spring of 1920.

days before the Göttingen Born–Jordan–Heisenberg paper was published in the German journal where Born was an editor! How on earth had that happened?

The answer was fairly simple. At the end of July, Heisenberg had gone to Cambridge to give a lecture, mainly on the pre-matrix mechanics of the atom. Ralph Fowler had a chat with him afterwards and learned a bit about Heisenberg's new idea for the atom, and a few weeks later he acquired proof sheets of his paper. He read it and then mailed it to Dirac in Bristol. At first Dirac did not think much of Heisenberg's work, but after a week or so he started thinking about it a lot.[22]

It wasn't until October that it finally hit him. He'd seen something like the commutator algebra before, and it was in the form of brackets that were a mathematical shorthand for the process of multiplying two quantities and then subtracting from the result the same two quantities multiplied in the reverse order. In the classical world the result was always zero, but Dirac could see that he could write a similar set of brackets for the same process in quantum physics. He promptly wrote up the paper; he sent a handwritten copy to Heisenberg and another went off for publication.

Heisenberg wrote back with disappointing news:

> Now I hope you are not disturbed by the fact that indeed parts of your results have already been found here some time ago and are published independently here in two papers—one by Born and Jordan, the other by Born, Jordan and me . . . On the other hand, your paper is also written really better and more concisely than our formulations given here.[23]

Dirac wasn't the only one working on completing the non-commutative algebra hinted at in Heisenberg's summertime paper. Kramers in Copenhagen and John Slater at MIT in the United States had both made the link with the bracket notation in classical physics and had adapted it for use in quantum physics. Slater had written up his paper and was about to send it off for publication when he found out that Dirac had beaten him to it.[24]

Undeterred, Dirac turned to the challenging task of seeing if the new matrix model could correctly produce the Balmer series for the hydrogen atom. By the end of November he knew that Pauli had already derived the hydrogen spectrum using matrix mechanics, because Heisenberg had sent him an early draft of Pauli's work, but he went ahead anyway and produced a more refined version. Moreover, Dirac's version was published first. Pauli's insistence that he not publish anything until it was exactly right meant he didn't publish nearly as much as other physicists, and sometimes, as now, others beat him to the punch. But Pauli was not as caught up as some scientists in the competition to see who could lay claim to an idea first.

Heisenberg was different. It might have provided Heisenberg with great satisfaction to return to Munich for Christmas as a success. After years of his father's anxious hand-wringing over whether Werner was really good enough to be successful in physics—and with a couple of embarrassingly forgettable atomic models behind him—he undoubtedly would have been delighted to be able to lay claim to the exciting new matrix mechanics. But Heisenberg wasn't sure about the new theory, his moods vacillating between hope and doubt.

Even Born was cautious about promising too much too soon. In the States, he was ending his MIT lectures with the proviso "This is only the first step towards the solution of the riddles of quantum theory."[25]

Both men were wise to be circumspect about the prospects of their new and untried model of the atom, since a very different way of looking at quantum physics was brewing in, of all places, an Alpine holiday lodge.

Chapter   Twelve

SHOCK   WAVES

The best laid schemes o' mice an' men gang aft agley,
An' lea'e us nought but grief an' pain, for promis'd joy.

ROBERT BURNS,
"To a Mouse"

Erwin Schrödinger's second preoccupation was women—physics
was the first—but he also had a strong philosophical bent. Before
heading off on his 1925 Christmas holiday in the Alps with his
amour du jour, he'd spent much of the fall writing a book of philosophy. It was a deeply personal account of his philosophy of life,
influenced by the holistic views of Ernst Mach and the Eastern
mysticism of the ancient Indian Vedanta. Biographer Walter Moore
has suggested that his intense focus on the big philosophical questions of life may have served as a subconscious marshalling force for
the creativity that would come next. "The testament of 1925 must
have been the result of many days of meditation; here I am, thirty-
eight years old, well past the age at which most great theoreticians
have made their major discoveries, holder of the chair that Einstein
once held, who am I, whence did I come, where am I going?"[1]
    It's not that Schrödinger hadn't done anything useful in his scien-
tific career. He'd published some decent papers, but his main problem was that he didn't stick with any particular field of study for very
long. By the fall of 1925 he'd published about forty papers, some
on mechanics, geophysics and colour theory, but most on atomic
physics and radiation theory. He'd become the acknowledged world

expert on colour theory in the early 1920s, and even produced his own "encyclopedia article," a 104-page paper on the mathematics of colour measurement and the physiology of the eye. Although his work outside colour theory was useful, it wasn't particularly dramatic. He was more likely, by his own admission, to react critically to others' scientific ideas and work on them than to propose new ideas of his own.[2]

Schrödinger's work on atomic theory was in the same vein as that of Einstein and Ehrenfest. It stemmed from Boltzmann's statistical thermodynamics rather than the spectroscopic fingerprints used by Sommerfeld, Bohr and Born at the Munich, Copenhagen and Göttingen schools. Schrödinger had the benefit of having worked in spectroscopy and knew the field quite well, so he understood both approaches to developing a theory of quantum physics without being particularly attached to either. Hence Schrödinger tended to be more open-minded about the weakness in spectroscopic research than its more committed adherents.[3]

The beginning of 1925 hadn't been an easy time. His year had started out with him and his wife arguing over whether to get a divorce. He and Anny had a somewhat unconventional marriage by societal standards, but not by the standards of their group of friends in Zurich. They all had affairs, often with each other's husbands and wives. It was all very chummy, but although Erwin thoroughly enjoyed his freedom to seek sexual satisfaction where it suited him, he wasn't sure he liked Anny enjoying her own liaisons quite so much. He had grown up surrounded by adoring women, and Anny wasn't being the doting wife he expected.

Erwin was a rather spoiled only child. His father, Rudolph, had reluctantly taken over running the family linoleum factory when he would sooner have been an artist or scientist, interests he pursued by studying Italian art and botany in his spare time. Rudolph's mother had died when he was only two, and his father had raised him and his brother on his own. Rudolph's strong relationship with his father was reflected in the bond with his own son, Erwin.

Young Schrödinger was a bright and inquisitive child, and his father encouraged him to explore a broad range of subjects. Although Erwin exhibited an early talent for mathematics, he also developed a strong interest in languages, poetry and the theatre. His mother, Georgie, was one of three pampered daughters of a chemistry professor who was very much involved in Austria's theatre scene. Like Rudolph's mother, Georgie's mother had died in childbirth when she was quite young, and her father had raised her and her sisters on his own.

Georgie's father owned a grand five-storey townhouse in Vienna, and rented the top-floor apartment to Rudolph and Georgie. There were always lots of women about the place to fuss over little Erwin—his delicate mother, her doting sisters, as well as maids and nurses. Erwin never wanted for feminine adulation, and perhaps came to expect it as his due.[4]

Erwin did very well in school, usually taking top honours in his classes. As an adolescent, however, he did not make male friends easily. On the other hand, girls fascinated Erwin, and he was constantly falling in love with the sisters of his schoolmates. It wasn't long before Erwin started keeping a written record of his loves, coded to indicate if or when he'd had sex with them.[5]

Schrödinger headed off to the University of Vienna in the fall of 1906 and immediately demonstrated an exceptional talent for mathematics and physics. He fully intended to join the many students who had studied with Ludwig Boltzmann, but it was not to be: Boltzmann hanged himself just a few months before Schrödinger was to begin working with the great master of statistical mechanics.

Boltzmann's death was a tremendous blow to the Vienna physics department, and it took about eighteen months of administrative squabbling before a replacement was finally appointed to fill his vacant chair. Fritz Hasenöhrl, one of Boltzmann's former students, proved to be an excellent choice, at least for Schrödinger. Inspired by the new, young professor, Schrödinger vowed to become a mathematical physicist and follow in the footsteps of Boltzmann and Hasenöhrl.

Boltzmann, Mach and their contentious relationship was still a big topic of discussion during Schrödinger's university years, but Schrödinger and his fellow students found a way to balance the conflicting realist and positivist philosophies of the two men. He later wrote, "We decided for ourselves that these were just different methods of attack, and that one was quite permitted to follow one or the other provided one did not lose sight of the important principles . . . of the other one."[6] Even in his student days Schrödinger exhibited a pragmatism that allowed him to appreciate multiple viewpoints without becoming overly attached to any.

Schrödinger's university training was not just in theory and mathematics. He also served as an assistant to one of the university's top physics experimentalists. He quickly found out he wasn't particularly talented in the lab, but he did develop an appreciation for the value of direct observation of laboratory measurements as a solid basis for supporting a theory.

He applied for his habilitation in the winter of 1913, but his performance had not been impressive. Like Heisenberg's dismal oral presentation for his doctoral degree, Schrödinger's caused considerable debate amongst academics and government officials. His habilitation was finally confirmed in January 1914, but it was a less than spectacular start to an academic career.

Schrödinger's lacklustre performance during his final years of study may very well have been due to a troubled romance. He had fallen madly in love with the daughter of one of his parents' friends. Felicie was much younger than Erwin; when they became romantically involved, he was twenty-three and Felicie fifteen. Felicie's class-conscious parents, who claimed a connection to minor nobility, were adamantly opposed to their daughter's relationship with the son of a linoleum manufacturer. Erwin would not be able to support their daughter in the style to which she was accustomed on the pitiful pay of a beginning lecturer. And he didn't even have a title of any kind! Desperate for a solution, Erwin tried to persuade his father to let him quit university and work at the linoleum factory so he could support a wife, but Rudolph wisely refused. He wanted his son to have the scientific career he himself had sacrificed for the family business. In the middle of 1913, Felicie, then seventeen, finally broke their informal engagement, leaving Erwin heartbroken.[7]

Biographer Walter Moore has speculated that Schrödinger's first truly deep love and devotion for a young woman was thwarted by the "bourgeois" requirements of a socially acceptable marriage, and that this soured his view of marriage. "Since he was prevented from a dedication of his spiritual, romantic and sexual longings to the one person of his choice, he would henceforth look with disdain upon the institution of marriage and attempt to construct his emotional life outside its rigid framework."[8]

Like de Broglie, Schrödinger was old enough to be called up for war service in 1914, just after he'd finished university. But unlike de Broglie, he did not escape active combat duty. Schrödinger spent most of the war as an artillery officer on the Italian front, with its periods of calm punctuated by bouts of bloody and brutal fighting. Much to Schrödinger's dismay, his adored professor Hasenöhrl was killed in battle.

In 1917, Schrödinger was sent from the front to teach science to soldiers in Vienna. Throughout his years in the military Schrödinger had found time to keep doing research, and he had studied Einstein's new theory of general relativity and Bohr's quantized atom. Finally, at the age of thirty, he got his career back on track and started publishing papers, producing his first essay on the thermodynamics of the quantum.

The shadow of the war could not be escaped even in Vienna. With the rationing of gas, the Schrödingers' fifth-floor apartment was often dark and cold, and there was a severe shortage of food. Life was grim. Georgie had undergone radical surgery for breast cancer, and her recovery was slow and painful. Erwin contracted a lung infection that may well have been tuberculosis, a disease sweeping the impoverished city. And then Rudolph finally shut down the factory because he could no longer get the supplies needed for making linoleum. During the winter of 1917 the Schrödingers found themselves regular visitors at a local soup kitchen.

When the war ended, times remained just as difficult in Vienna as in Berlin or Frankfurt. Erwin sought refuge at the university, writing papers on colour theory and publishing another on wave/particle duality in which he proposed a possible experiment to determine if light was a wave or a particle. His own painstaking experiment in a dark and cold university laboratory had convinced him of the wave nature of light.

The privations soon began taking their toll on Erwin's father. Rudolph had neither the strength nor the interest to reopen the linoleum factory in the aftermath of the war, and the family had lost its savings in useless war bonds. Rudolph died quietly, resting peacefully in a big chair in their fifth-floor apartment, on Christmas Eve, 1919.

Erwin and his mother were left with nothing to live on but Erwin's pittance for lecturing, so Georgie's father asked them to vacate the apartment; he needed the rental income if he was to survive himself. Georgie went to stay with her comfortably well-off sisters, but Erwin was now without a home. He needed to secure a position as a university professor if he was to be able to have a decent home and to support his latest love, Anny.

Anny was a farm girl from Salzburg who had moved to Vienna to work as a secretary, earning more money in a month than Erwin made in an entire year. They wanted to marry, but Erwin first needed a decent job so that Anny wouldn't have to keep working. Erwin had a particular fondness for girlish, much younger women, but Anny was already in her early twenties when they talked of marriage, and she was more tomboyish than delicately feminine. Perhaps he looked past her age and sturdy build because of her unstinting, fervent adoration for him. And besides, he wanted a home. It had been quite a while since Erwin had been able to rest in the warm and nurturing bosom of unquestioning worship.

Erwin did finally obtain a professor's position in 1920, and he and Anny got married. But the job was only a short appointment in Jena, and it was followed by another in Stuttgart, and yet another short-term job in Breslau. Unfortunately, Georgie's cancer returned, and she died in the fall of 1921. But Erwin's dark days were nearly over. Almost immediately after his mother's funeral, Schrödinger got an offer of the position as chair of theoretical physics at the University of Zurich. It was the offer he'd been waiting for, and once more Erwin and Anny packed up and moved to a new home.

Despite the war, Schrödinger had been doing research in physics for nearly a decade, and finally, at age thirty-four, he had an excellent professorship at a prestigious university. In all those years he'd got to know some of the key players in European physics. He could hardly forget the day just after his graduation when an enthusiastic Ehrenfest had dragged him off to a Viennese coffee house to expound on the delights of physics. Others Schrödinger had met at conferences and lectures. In particular, he'd been inspired by Einstein's lecture on the problems of gravitation at a huge conference in Vienna in 1913. The who's who of physics had been there, as well as some of the younger scientists such as Max Born, and the place had been abuzz with talk of the new quantized atom published by a young Danish physicist.

When Schrödinger started his new job in Zurich in the fall of 1921, he was emotionally and physically exhausted. By November he'd come down with another severe respiratory ailment and was ordered to take a complete cure at a high altitude, then considered the best environment for curing tuberculosis and other breathing troubles. Erwin and Anny took up residence in the Alpine resort of Arosa. His cure took nine months, but he didn't forget about physics. He considered the problems with quantized orbits and the Bohr atom during his convalescence, and wrote up a short note hinting at the wave nature of atomic orbits.

Schrödinger returned to work in the fall of 1922, and in November he wrote to Wolfgang Pauli, then in Copenhagen, about his ideas for his inaugural lecture. Schrödinger, like others who were struggling with the puzzle of atomic radiation, had begun to wonder if conservation of energy might have to be considered statistical. After all, probabilities were considered acceptable for calculating entropy under the second law of thermodynamics. "I for my part believe, *horribile dictu*, that the energy-momentum law is violated in the process of radiation . . . Can the conservation of energy-momentum not be merely a macroscopically valid average relation, of which atomic physics knows nothing, like the 2nd Law? At least it can be that way, and I see almost no other way out."9

His lecture, on December 9, 1922, was called "What is a Natural Law?" In it, Schrödinger attempted to clarify the probabilistic nature of laws. "If the probability that an event occurs in a given way is so high that from the human viewpoint it has become a certainty, then we speak of a 'law of Nature.' This is possible, however, only if there is an unimaginably large number of individual occurrences, as can happen in a molecular process."10 Schrödinger's argument very much mirrored that given by Boltzmann when he insisted that the irreversibility of entropy was not a certainty but just seemed that way because the probability of it being otherwise was so tiny. But Schrödinger also added an argument for acausality to his lecture. Since the large-scale laws like conservation of energy, which covered enormous ensembles of molecules, could be considered statistical, he said, it made no sense to regard individual events as anything but random, or acausal. When the Bohr, Kramers and Slater paper came out of Copenhagen some two years later, it's not surprising that Schrödinger was immediately enthusiastic about it; it contained the very ideas he had discussed with Pauli and espoused in his inaugural lecture.

Schrödinger's next boost up the academic ladder came in the spring of 1924 when he was invited to attend the Fourth Solvay Conference as an observer. He wouldn't be presenting a paper, but the invitation helped to establish him as a scientist to take seriously. It also inspired him to get going on his lagging research, which he did with enthusiasm, publishing six papers in 1924. One tackled a model of the quantized hydrogen atom, where Schrödinger was able to predict energy levels that matched experimental data, although it soon turned out to be flawed.

In the fall, Schrödinger attended a physics conference in Innsbruck that attracted a large number of physicists, including Einstein, Sommerfeld, Planck, Born and Pauli. It was there that Pauli finally got Einstein's decidedly negative opinion on the BKS paper, which Pauli then related to Bohr at Copenhagen. Schrödinger didn't present a paper himself, but he participated enthusiastically in the discussions and enjoyed renewing acquaintance with other scientists.[11]

For their Christmas holidays, Erwin and Anny went back to Arosa, just as they'd done every Christmas since Erwin's illness in 1921. But there was a new tension in their relationship. Erwin and Anny had plenty of romantic drama in their lives, just not with each other; and Erwin couldn't help but be annoyed by the fact that Anny was not the least bit concerned about his various liaisons. It was a double standard, of course, but he seemed to expect Anny to be accepting of his falling in love with any attractive woman who walked by while at the same time wanting her to be at least somewhat jealous of his attentions to other women.

Even so, Erwin didn't want a divorce, ever mindful of the cost involved. His soup kitchen days weren't so far behind him that he could be reckless about money. Ultimately, they came to an arrangement that worked for them both: they would remain married, but each would be free to pursue sexual satisfaction elsewhere. It was a marriage of convenience, rather like that of Albert and Elsa Einstein (although it's not clear that either Albert or his wife ever considered that Elsa should be allowed the same freedom to take lovers that her husband exercised).

Early in 1925, Schrödinger read Einstein's January papers on Bose gas statistics, since statistical mechanics was one of his areas of interest. Following up on Einstein's reference to de Broglie in those papers, Schrödinger sought out the Frenchman's thesis to read it for himself, but he didn't get hold of a copy until the fall. It immediately struck a chord, because he'd had some similar ideas of his own that he'd published in a note in 1922. "As far as I can see," Schrödinger wrote to Einstein in early November 1925, "the mathematical content is the same, only mine is much more formal, less elegant, and not actually shown in general. Naturally, de Broglie's consideration in the framework of his comprehensive theory is altogether of far greater value than my single statement, which I did not know what to make of at first."[12]

De Broglie imagined an electron not as an object but as a wave vibrating around the nucleus with an integer number of wavelengths, or "beats." Think back to the wave travelling along a rope attached to a post. Now take away the post and attach the ends of the rope so that it forms a ring. Imagine that the rope is still moving up and down in a wave pattern, but the wave is no longer travelling up the rope and back again. The ring is essentially moving up and down, undulating in place as a standing wave—but this will work only if the ring has whole-number wavelengths. It can have any number of wavelengths, depending on how energetic the wave is, but it must always have a full number. If it had half or one-seventeenth of a wave, the awkward undulations would make the whole thing collapse. Thus, de Broglie could take the known allowed energies for quantized orbits, divide the energy by Planck's *h* and produce the proper wavelengths for each orbit. And since a wavelength is proportional to the inverse of a frequency, it was a simple step to match de Broglie's wavelengths to frequencies of the spectral lines.

A few weeks later, Schrödinger gave a lecture on de Broglie matter waves to the Viennese physics students. Felix Bloch was one of the students at the lecture. "Schrödinger gave a beautifully clear account of how de Broglie associated a wave with a particle and how he could obtain the quantization rules of Niels Bohr and Sommerfeld by demanding that an integer number of waves should be fitted along a stationary orbit."[13] According to Bloch, another professor then pointed out that if one wanted to take atomic waves seriously, one needed a wave equation. Schrödinger got working on it right away, but it didn't work out as easily as he'd hoped.

On December 15, Schrödinger sent off another paper, "On Einstein's Gas Theory," for publication. In it, he clearly stated his support of matter waves: "This means nothing else but taking seriously the de Broglie–Einstein wave theory of a moving particle, in which the particle is nothing more than a kind of a 'wave crest' on a background of waves."[14]

But he wasn't making much progress with his own wave equation. He'd tried using the relativistic theory from de Broglie's thesis, but the solutions were not right. So he put his notes away and headed back to the Alpine retreat in Arosa for his usual Christmas vacation. There is no doubt that Schrödinger had matter waves on his mind, but he also had a woman on his arm. This time Anny did not accompany him to Arosa. Instead, he brought one of his lovers.

However it happened, by the time he returned to Vienna from his Christmas tryst, he had worked out a non-relativistic wave equation.

As 1926 began, the scientists at Göttingen and Copenhagen were hopeful that matrix mechanics was indeed the much-needed breakthrough they'd been waiting for. With both Pauli and Dirac verifying that the unwieldy mathematics did indeed correctly predict the spectrum for the hydrogen atom, it seemed that, at last, quantum physics was emerging from the muddle and confusion that had bedevilled it for so long.

On January 27, 1926, Bohr wrote to Ernest Rutherford that perhaps the "grave difficulties in the spectral theory" had at last been overcome with the "spin" that the Leiden students had come up with. And he confided that he had been finding it very difficult to provide an "unambiguous use of mechanical pictures" in describing the quantized atom. Perhaps at last there was a light at the end of the tunnel. In his own convoluted way, Bohr gave credit to Heisenberg for the breakthrough. "In fact, due to the last work of Heisenberg prospects have with a stroke been realized, which although only vague[ly] grasped have for a long time been at the centre of our wishes."[15]

The "last work of Heisenberg" was, of course, the work of Born, Jordan and Heisenberg. However, Bohr was inclined to see Heisenberg's breakthrough as a product of his time in Copenhagen and the influence of both Kramers and Bohr. Ironically, Heisenberg, with his careless disregard for acknowledging the work of others, had not bothered to cite Bohr's work in his observables-only summer paper, even though he had used Bohr's correspondence principle. Bohr noticed the oversight.[16]

Not everyone amongst the small number of physicists who knew about the new quantum mechanics was as hopeful as Bohr. Whatever had transpired between Bohr and Einstein during their stay at Ehrenfest's house, they did not emerge from their discussions with a shared view of matrix mechanics. To Einstein, it was more like mathematical trickery that divorced itself from the classical Cartesian space-time coordinates. "The most interesting recent theoretical achievement," he wrote to his long-time friend Michele Besso, "is the Heisenberg–Born–Jordan theory of quantum states. A real sorcerer's multiplication table, in which infinite determinants (matrices) replace Cartesian coordinates. It is extremely ingenious, and thanks to its great complication sufficiently protected against disproof."[17]

In Bohr's lengthy letter to Rutherford, he made no mention of either Born or Jordan. Perhaps he merely used "the work of

Heisenberg" as shorthand for "the work of Heisenberg, Born and Jordan." Or maybe Bohr preferred to believe that Heisenberg's efforts had been inspired by the time he'd spent at Copenhagen just prior to his summer paper, and that the Göttingen crew had contributed only a few small items from the mathematical tool box. Hence, matrix mechanics was a Heisenberg/Copenhagen production.

On the same day that Bohr wrote to Rutherford, Schrödinger's paper on his new wave mechanics, incorporating the wave equation he'd derived on his Christmas holiday, arrived at *Annalen der Physik*. Schrödinger had worked alone during the three weeks since he'd returned to Vienna after his vacation. He'd found a non-relativistic wave equation that would give the correct energies for the quantized hydrogen atom, and he'd done it without the discontinuities that Born had built into matrix mechanics *and* without the quantum jumps that Schrödinger had always despised. In fact, he'd done it without matrix algebra at all, employing instead the comfortably familiar (at least to physicists) differential equations that were used in describing the motion of waves, whether they were water waves, sound waves or radiation waves.

Unaware of what Schrödinger was up to in Vienna, Max Born gave his final Boston lecture on quantum physics on January 22, 1926, before a crowd of one thousand people. His stay at MIT had been a great success, scientifically. The notes from his lectures were to be printed in book form and made available to American physicists and students. Born had also found a collaborator not long after he and Hedi had arrived, a young American mathematician who had studied with Hilbert and Courant at Göttingen. They produced a paper on what happens when atoms collide, which could not be determined using the discontinuous mathematics of matrix mechanics, so there were some problems to overcome. Nonetheless, they went as far as they could before the Christmas holidays. It was the first paper on the new quantum physics to be written in the United States.[18]

On the downside, Hedi was still feeling ill from her recent bout of food poisoning. Rather than continue with the American tour, it seemed wiser for her to return to Germany and take up residence at a prestigious sanatorium in Frankfurt to take "the cure."[19] Born continued his tour at the end of January, albeit disappointed that Hedi could not be with him. He would be preaching the Göttingen gospel of matrix mechanics at about a dozen major universities and research centres across the country, from Harvard to Berkeley. So many American scientists and physics students were eager to devour the newest theory in quantum physics—a Göttingen success story.

In February, back in Göttingen, Heisenberg and Jordan were busy putting the icing on the matrix mechanics cake. They wanted to integrate the new "spin" into the matrix model, but Pauli was still opposed to the idea. In early March, Pauli went to Copenhagen on his Easter holidays, and there spent some time with experimentalist Llewellyn Thomas, one of Ralph Fowler's students from Cambridge. Thomas had just effectively tidied up any last remaining doubt about the existence of spin. It still took another week or two for Pauli to climb down from his perch. He really, really did not like being wrong.

Thomas wrote, in a rather gleeful note to Samuel Goudsmit in Leiden, that it was just as well that Goudsmit and Uhlenbeck had published their note on spin before Pauli got to them. Pauli, the "whip of God," had been mistaken, and it was rare to see it happen so publicly. "Which all goes to show," said Thomas, "that the infallibility of the Deity does not extend to his self-styled vicar on earth."[20]

Within days of Pauli's capitulation on spin, Jordan and Heisenberg reworked the paper they'd just about finished to incorporate the subtleties of the magnetic field caused by the spin of the electron. The fine-tuned matrix mechanics had the blessing of Bohr and Pauli. It wasn't the final word, of course, because they still hadn't figured out what to do with Pauli's electron exclusion, but they were hopeful it would be solved before too long. For now, it was cause for celebration that matrix mechanics was proving to be such a good model for atomic physics. Granted, the mathematics was daunting, but physicists were just going to have to learn the complexities of matrix algebra if they were going to do the new atomic physics.

Pauli, however, was aware that something new was in the offing. Willy Wien, an experimentalist at Munich and the editor of *Annalen der Physik*, was the first to read Schrödinger's manuscript introducing wave theory, which landed on his desk on January 27, 1926. Schrödinger had asked Wien to give his papers to Sommerfeld to read before they were published, since he valued Sommerfeld's opinion. Sommerfeld had the opportunity to read Schrödinger's first two papers, the second arriving on Wien's desk on February 23, before he headed to England to give some lectures. Sommerfeld let Pauli know that it would be a good idea to keep an eye out for a soon-to-be-published "crazy" new theory.[21]

As soon as Sommerfeld had read the first two Schrödinger papers, he raised an obvious question: were wave mechanics and matrix mechanics somehow related? Schrödinger couldn't immediately see the relationship, even though he was convinced it existed. "Consequently," Schrödinger wrote to Wien, "I have given up looking any further myself. This I have done more happily since the matrix calculus was already unbearable to me long before I ever distantly thought of my theory . . . Now I firmly hope, of course, that the matrix method, after its valuable results have been absorbed by the [wave] theory, will disappear again."[22]

Pauli, still in Copenhagen, did indeed keep an eagle eye out for the new *Annalen der Physik*, and as soon as it arrived, at the end of March, he immediately got to work assessing Schrödinger's new theory. He was impressed by the fact that Schrödinger had derived the Balmer series for the hydrogen atom in a fairly simple manner, something Pauli could well appreciate since it wasn't that long ago that he'd derived the same thing using the tortuous mathematics of matrix mechanics. Pauli sent Jordan a long letter detailing his evaluation of the new wave mechanics, advising Jordan to take it seriously. "I feel this paper is to be considered among the most important recent publications. Please read it carefully and with devotion."[23]

Bohr was not quite sure what to make of it all. "[Pauli] is fascinated by the just published paper by Schrödinger," Bohr wrote to Heisenberg, who was still at Göttingen, "and has been able to prove that Schrödinger's scheme leads to results which agree with those obtained from your quantum mechanics."[24]

Although it was beginning to look as if there was a serious new competitor on the scene to challenge the Göttingen matrix model, Born was not particularly perturbed. After spreading the Göttingen gospel in the United States, it would not have been surprising if he was devastated when he found out about Schrödinger's wave mechanics on his return to Germany at the end of March. Instead, what upset him was discovering Jordan's forgotten paper in his suitcase. Jordan had used Pauli's electron exclusiveness to write a new probability equation that accounted for the severely restricted ways that electrons could arrange themselves. Unfortunately, Enrico Fermi, a young scientist from Italy who had spent much of 1924 in Göttingen and Leiden, had done the exact same thing, and his paper had been published in February. It was too late to submit Jordan's paper, and Born was appalled to find that he had inadvertently prevented Jordan from being the first to publish the new statistics and therefore getting the credit. It was a missed opportunity that Born deeply regretted, although Jordan himself didn't seem to harbour any hard feelings over the incident.

However badly Born felt about forgetting Jordan's paper, he was not at all dismayed by the new quantum wave theory. It was based on the continuous mathematics that were just what he needed to improve on the atomic collisions paper he'd written when he was at MIT. Besides, Born had already made himself familiar with the de Broglie matter waves that had been the genesis for wave mechanics.

If Born was disappointed about anything to do with Schrödinger's waves, it was the fact that he now realized he might very well have come up with wave mechanics himself if he hadn't dropped it the previous summer to attend to Heisenberg's paper.[25] But rather than dwell on what might have been, Born got to work on a new collision paper, this time using the new wave mechanics.

At Cambridge, Dirac was focused on writing his doctoral thesis and found the idea of a whole new theory to be really annoying. "Why should one go back to the pre-Heisenberg stage when we did not have quantum mechanics and try to build anew?"[26] Dirac was rather peeved that another theory had come along when they had a perfectly good one already. He just didn't see the point of it.

Einstein, who had never been impressed by the "sorcerer's" mathematics of matrix mechanics, wrote to Schrödinger with warm praise, saying "the idea of your work springs from true genius!"[27]

Ehrenfest was equally delighted. He was "simply fascinated" by the new theory, he wrote to Schrödinger with his usual dramatic flair. "Every day for the past two weeks our little group has been standing for hours at a time in front of the blackboard in order to train itself in all the splendid ramifications."[28] The chalk dust was flying.

For Heisenberg, the turnaround was stunning. Only a few weeks ago he had been the wunderkind at the forefront of quantum physics. Suddenly, out of nowhere, Schrödinger had single-handedly produced a completely new theory that was capturing enthusiastic accolades from the same physics community that had, so far, pretty much ignored his matrix mechanics.

Heisenberg's thunder was being stolen, and, by association, so was Bohr's, since Heisenberg's matrix mechanics had become a Copenhagen creation (however much Born and Jordan might consider it a Göttingen effort). But Heisenberg was not about to let his ambitions be thwarted so easily. He was prepared to do battle for the supremacy of matrix mechanics, which, of course, was much the same thing as doing battle for the supremacy of himself.

Chapter Thirteen

DRAWING THE BATTLE LINES

It is better to debate a question without settling it
than to settle a question without debating it.

JOSEPH JOUBERT, French philosopher

The name-calling started almost immediately. For Heisenberg, Schrödinger's wave mechanics was "bullshit," "abominable" and "repelling." Indeed, Heisenberg considered all papers that made use of wave mechanics to be repellent. Matrix mechanics was obviously superior, he said, because it better represented the undeniable discontinuous nature of the quantum world. In the category of damned-by-faint-praise, Heisenberg conceded that wave mechanics, being devoid of any physical meaning, was at least a helpful mathematical tool—for calculating matrices.[1]

Initially, Schrödinger had no quarrel with Heisenberg or matrix mechanics, even if he didn't care for its mathematical complexity. Schrödinger started out with a mathematical model in his January paper, hoping to later develop a physical meaning for the symbol ψ (which he used to represent a matter wave) so that it "would more nearly approach reality than the electron orbits, the real existence of which is being very much questioned today."[2] Schrödinger wasn't insisting on going back to the old model of the atom that had electrons orbiting the atomic nucleus. He was onside with Heisenberg in questioning the reality of electron orbits.

Visualizability was fundamental to Schrödinger for his understanding of physics, but he did not mistake the use of a good
metaphor for reality. In his own particular blend of the philosophies
of both Ludwig Boltzmann and Ernst Mach—with a little Eastern
mysticism thrown in—he firmly believed in the intelligibility of the
universe and the creation of a unified world view through the use of
visualizable metaphors or analogies. Such images were useful tools
for developing a physics model, but Schrödinger did not necessarily
believe they represented a real world.[3]

The Zurich physicist had not developed his wave theory in
opposition to matrix mechanics, but rather as an exploration of the
wave nature of the quantum world. But Heisenberg couldn't help
but see it as a very personal threat, and his reaction to Schrödinger's
theory was immediately hostile. He hoped it was simply wrong, and
in some ways it was, because it was still a work in progress. But then,
so was matrix mechanics.

The wave and matrix mechanics had been birthed in a similar
fashion. Some scientists, such as Einstein and Pauli, did not publish
a theory until it was completely thought out and the flaws and con-
tradictions were dealt with. That's why it took Einstein nearly a
decade to develop his theory of general relativity, and why Einstein's
1905 papers were continuing to be used just as they were originally
written; they were complete and correct when they were published.
Pauli wrote extensively about his ideas in letters to his colleagues,
which they then copied and shared about with whoever wanted to
study and use them. Pauli worked out his ideas so carefully that
some of his letters could have been published in scientific journals,
but neither he nor Einstein cared about being first. They cared
about being right.

The matrix mechanics papers that had appeared in the summer
and fall of 1925 had been produced in a rushed and piecemeal fashion. First came Heisenberg's observables-only summer paper. This
was followed by the Born and Jordan paper, which made explicit
use of matrices. Indeed, Born may well have rushed the second

paper so that he—and Göttingen—would be first to lay claim to the matrix model. This was followed by a third paper by all three that more broadly fleshed out the new matrix mechanics, but it had been prepared under the pressure of Born's imminent departure for the United States.

Schrödinger's process was much the same except that he worked alone. From time to time he did consult on the mathematics with his Zurich colleague Hermann Weyl, who was at that point embroiled in a passionate love affair with Schrödinger's wife, Anny. Since the Weyls were part of the same sexually permissive crowd as the Schrödingers, the affair was no cause for tension between the two colleagues. Schrödinger's first paper on wave mechanics was submitted for publication at the end of January 1926; the second, more detailed version in February; with the third paper, on the equivalence of the matrix and wave mechanics, in March. His fourth paper wasn't submitted until the end of June because he had first to address the problems in the earlier papers.

Initially, Schrödinger had hoped his wave function ψ represented a real matter wave in real space. He imagined his wave function as describing the characteristic vibrations that matched the allowed energy states of the atom. What had once been the space that electrons had to "jump" between to get from one allowed orbit to another in the Bohr atomic model was, in wave mechanics, replaced by energy passing from one vibration pattern to another. There was no jumping going on at all. Schrödinger envisaged the radiation of light as a kind of "beat" caused by the transition from one set of vibrations to another, with transitions happening continuously in space and time.[4] Thus, Schrödinger could retain continuity in classical space and time while eliminating Bohr's old discrete stationary (allowed) orbits and "impossible quantum jumps."

Schrödinger had never liked quantum jumps since they made it impossible to visualize what was going on inside the atom, and it didn't make any sense to him to replace visualizable space-time physics with non-visualizable mathematical abstraction if one wanted to understand physics. He didn't buy the premise of matrix mechanics which stated that one could never know the real structure of the atom. "For we cannot really avoid our thinking in terms of space and time," he wrote in his second wave theory paper in February, "and what we cannot comprehend within it, we cannot comprehend at all. There *are* such things but I do not believe atomic structure is one of them."[5]

Schrödinger's third paper in March resolved the question Arnold Sommerfeld had asked about the relationship between wave and matrix mechanics. The answer? The two theories were just using two different approaches to arrive at the same solutions to the description of the quantized atom. They were equivalent, which meant that busy physicists would not have to bother learning complicated matrix algebra if familiar differential equations would do the same job.

Wolfgang Pauli had already demonstrated that fact even before Schrödinger's third paper came out. In April he'd sent a letter to Pascual Jordan containing a lengthy analysis of wave mechanics, in which he concluded that wave and matrix mechanics were equivalent. That left Heisenberg in a difficult spot. He couldn't argue anymore that wave mechanics was wrong because, by equivalence, matrix mechanics would have to be wrong too. Instead, he would have to show that his theory was superior to Schrödinger's, and that was going to be a tough sell to physicists who were already inclined to favour wave mechanics because the mathematics was familiar and so much simpler than matrix algebra.

Heisenberg had more than just his ego on the line. In the spring of 1926 he'd been offered a full professorship at the University of Leipzig. German academic tradition dictated that a recently (or soon-to-be) habilitated young scientist had to accept the first offer of a professorial appointment or risk being sidelined in the competition for the limited number of good jobs. But Heisenberg had already agreed to go to Copenhagen to replace Kramers as Bohr's right-hand man. After consulting with a number of people, including Born and Courant at Göttingen and Einstein in Berlin—who all encouraged him to go to Copenhagen—Heisenberg decided to take the career risk. August Heisenberg thought it was a bad idea. It made no sense to him for his son to give up the much-coveted position of professor in Germany—a position he himself had worked so hard to obtain—for the questionable rewards of a short-term assistantship in Denmark. He was worried that Werner had made a big mistake.[6]

At the time of the Leipzig offer, Heisenberg was riding high on the success of matrix mechanics, and he was confident there would be more German professorships on offer. But once it became apparent that matrix and wave mechanics were equivalent, he could no longer count on his work on matrix mechanics being enough to bring another job offer. He had to make sure that his theory was acknowledged as superior and that he was still the wunderkind. It wasn't as if many people were using matrix mechanics, other than its inventors and their assistants, and maybe a few other physicists such as Pauli and Dirac. Even Arnold Sommerfeld was reluctant to use it, since he considered matrix mechanics to be "frighteningly abstract."[7]

Felix Bloch, a physics student at Zurich, attended Schrödinger's lectures on the new wave mechanics in early 1926, and followed his papers as they appeared. "I did not learn about the matrix formulation of quantum mechanics by Heisenberg, Born and Pascual Jordan until I read that paper of Schrödinger's in which he showed the two formulations to lead to the same results."[8] Indeed, Schrödinger's equivalence paper gave matrix mechanics a credibility in the physics community it had not had before. Matrix mechanics, by itself, could have been just one more daft idea being floated in an attempt to explain quantum phenomena. But if matrix and wave mechanics both produced the same correct solutions, maybe the matrix method had something going for it after all.

But wave mechanics swept past the matrix variety so quickly that physics students didn't have to struggle with matrix algebra; they went directly to wave equations. And so did the growing number of physicists working on the quantized atom.

The appearance of two theories of the quantized atom caused a wave of excitement in the physics community. A colloquium at the University of Berlin in the spring of 1926 on the topic of the two new theories attracted a big crowd. About two hundred physicists and students crammed into a hall to hear about the new models. Einstein was there as well. One physicist who attended the meeting recalled how it unfolded.

> After the report, Einstein rose and said, "Now listen! Up to now we had no exact quantum theory, and today we suddenly have two of them . . . You will agree with me that these two theories exclude one another. Which theory is the correct one? Perhaps neither of them is correct." At this moment—and I shall never forget it—[another physicist] stood up and said, "I have just arrived from [Hamburg]. Pauli has proved that both theories are identical." . . . This was a wonderful moment to see that two theories, apparently based on totally different fundamental ideas, nevertheless both describe nature accurately. I feel this was the birthday of modern quantum theory.[9]

Towards the end of May, Heisenberg wrote to Dirac at Cambridge and explained Pauli's equivalence analysis, but reiterated his criticism of Schrödinger's theory of matter waves since it "must be inconsistent as just like the wave theory of light."[10] Dirac promptly dropped his opposition to wave mechanics—which was probably not what Heisenberg intended—because he realized that it was just common sense to use wave mechanics if it made computations easier. With his background in engineering and mathematics, Dirac was not familiar with classical wave theory, but he immediately set to work getting up to speed on it. And as far as the interpretation went, Dirac didn't really care if Schrödinger's ψ was real or not; that was metaphysics, not science.[11]

Dirac quickly came up with a paper that incorporated Pauli's electron exclusion into the calculation of the probability of a transition between two states. He was embarrassed, however, to find out that Enrico Fermi had already published a paper with the same results in February, and that Jordan had come up with the new statistics the previous fall (although his paper had remained hidden in the bottom of Born's suitcase for several months).

Pauli, Dirac, Jordan and Born all conceded that wave mechanics was a very useful addition to quantum physics. Even Bohr was receptive to the new wave theory. He was already looking for a theory that would incorporate what now appeared to be a fundamental principle of the quantum world: the wave-particle dual nature of reality. The presence of a particle theory (matrix mechanics) and a wave theory (wave mechanics) served only to reinforce Bohr's belief that he was on the right track.

Heisenberg had initially railed against Born for so quickly publishing the collision paper using wave mechanics, calling him a "traitor" to quantum mechanics. But as Schrödinger elaborated on the physical principles of wave mechanics, Born's alliance shifted back to matrix mechanics. Born had built discontinuities into his quantum mechanics from the start, which was, in his mind, a fundamental principle of the quantum world. But Schrödinger's theory was based on continuity, something Born could not support. Jordan joined Heisenberg and Born in defending the superiority of matrix mechanics. By the summer of 1926 the battle lines were becoming clearer.

It was also clear by the summer that both versions of quantum mechanics had serious problems. Because matrix mechanics precluded visualization, it was difficult to see how to advance it other than by plucking something new from the mathematical tool box and hoping it might produce a useful result. It was, after all, essentially the method Born had used to shore up Heisenberg's summer paper with its strange multiplication equation.

An even bigger problem was the inability of matrix mechanics to solve outstanding problems such as the intensity of hydrogen spectral lines, or to come up with the correct energies for the helium atom. These issues were no more solvable than they had been under the old quantum physics. And, despite Bohr's correspondence principle being part of Heisenberg's reasoning in his summer paper, matrix mechanics itself did not offer a means of making a transition from the quantum world to the classical world.[12]

Schrödinger, defending his version of quantum mechanics, answered Heisenberg's name-calling with his own. He called matrix mechanics "repelling," "irrational," even "horrifying." To give up visualization and the classical space-time description was the same as complete surrender, charged Schrödinger, and its very lack of visualizability made matrix mechanics useless for advancing science.[13] Yet wave mechanics had had its own evolutionary problems. Schrödinger hadn't been able to find any way to get his matter waves to interact with electromagnetic radiation waves, so his interpretation of the matter wave ψ needed reworking. Schrödinger had also proposed, in a blending of wave-particle duality, that electrons were actually tiny "wave packets." Hendrik Lorentz, who had been carefully following the development of wave mechanics, pointed out that the "wave packet" electrons would inevitably lose energy and spread out over time, as waves were wont to do, so that they wouldn't be packets anymore.

Schrödinger let go of the idea of electrons as "wave packets," allowed that his function ψ, when squared, represented all the possible states a wave could be in, but included imaginary space* as well as real space. His wave function could have an imaginary part as well as a real part. Schrödinger's revised wave mechanics did away with ψ as an actual matter wave, since it no longer existed in three-dimensional space but rather in configuration space, a "space" where

---

*Imaginary space is a mathematical concept. It is just like regular space except it allows that there is a square root of minus one.

the number of "dimensions" matched the number of ways an entity could move.* This was a significant step back from his ideal of painting a nice picture in classical space. Instead, the new wave function could be considered a superposition of one or even all the possible allowed states, existing all at the same time. In other words, every allowed vibration for an atom could be present in the wave function. "If one likes paradoxes," wrote Schrödinger, "one can say that the system is found simultaneously in all conceivable [states] but not in all of them in equal strength. . ."[14] Schrödinger was still advocating a visualizable model for the quantized atom, but the idea that had once been based on a simple matter wave vibrating around a nucleus had been forced to evolve into a less classical and more abstract theory.

Physicists were eager to hear the details of the new wave mechanics, and as soon as Schrödinger finished his fourth paper in June, he was on the move. First up was a lecture at a physics conference, conveniently being held in Zurich at the end of June. Then there were several lectures in Germany. After the talk in Berlin, Max Planck threw Schrödinger a big party, and that's when Planck started giving serious consideration to Schrödinger as his possible replacement. He would be retiring from the chair of theoretical physics at the University of Berlin in October.[15] Munich was next. Sommerfeld had wanted Schrödinger to come to Munich in the spring to give a talk on wave mechanics, but Schrödinger had begged off. He'd been feeling under enormous strain, between developing wave mechanics and managing his students, especially since he didn't have an assistant at the time to help with the students.

Heisenberg, home for the summer with his parents, made a point of attending Schrödinger's lecture. In the question period

---

*In classical space, an object can move in only three ways (three dimensions): up/down, forward/back and side to side. In configuration space, the number of dimensions matches the number of ways an object can move (the degrees of freedom), which in some cases can be infinite.

following the talk, Heisenberg demanded to know how Schrödinger could hope to explain discontinuities such as the photoelectric effect when his model was based on continuity. As Heisenberg reported to Pauli afterwards, Willy Wien, the erstwhile editor of *Annalen der Physik* who had never liked matrix mechanics, angrily interrupted. "Professor Schrödinger will certainly take care of all these questions in due time. You must understand that we are now finished with all that nonsense about quantum jumps."[16] Wien may have been a little biased against Heisenberg. Wien had been an examiner for Heisenberg's doctoral defence, and he was the one who had wanted to deny the "prima donna" Heisenberg a doctoral degree.[17]

If Heisenberg had any doubts about how matrix mechanics stacked up against wave mechanics in the eyes of Europe's physicists, they were now gone. The gathering at the Munich lecture wouldn't even listen to his arguments for matrix mechanics. Heisenberg immediately wrote to Bohr about what was going on, and Bohr promptly wrote to Schrödinger to invite him to Copenhagen. Bohr wanted to hear about wave mechanics from the horse's mouth, so to speak.[18]

But Bohr was going to have to wait. Schrödinger had had a demanding year so far, and he needed a rest. His half-*annus mirabilis* had taken its toll. But he perked right up when he returned to Zurich. Anny had arranged for him to do some tutoring, which might not have gone down well with Erwin except that his students were the lovely twin daughters of a family friend, Withi and Ithi. They were just about to turn fifteen, an age Erwin found most attractive in nubile young women. As the summer progressed, he found it increasingly difficult to give his full attention to physics. He'd fallen in love with Ithi.

Heisenberg had no such tempting distractions to divert him from thinking about physics. Matrix mechanics needed a facelift if Heisenberg hoped to convince physicists that his mechanics was superior to Schrödinger's.

Despite Schrödinger's mixed bag of personal philosophies, he had positioned himself as more of a realist than not. He believed that physics, at the microscopic and macroscopic levels, was intelligible, otherwise there would be no point even bothering with the study of physics at all. Matrix mechanics, on the other hand, was embedded in positivism, whether Heisenberg, Born and Jordan had deliberately designed it that way or not. Logical positivism espoused the view that a mathematical truth was the same as a physical truth, and that metaphors or images were to be shunned. But Heisenberg had no particular allegiance to any philosophy, and when it started becoming apparent that the non-visualizable and highly abstract mathematics of matrix mechanics was not going over very well, he had no difficulty changing his position. So he did.

Quantum jumps had been eliminated from the Göttingen matrix mechanics from the start. Electron orbits were not observable, therefore the "jumps" between orbits were also not observable, and what was not observable would not be included in matrix mechanics. But, since wave mechanics didn't have quantum jumps, Heisenberg decided that matrix mechanics would. It turned out that matrix mechanics was quite unaffected by the return of electrons, electron orbits and quantum jumps. Leaving them out had merely been the philosophical preference of the Göttingen crew.

In between tutoring and cuddling with the teenaged Ithi, Schrödinger was still giving some of his attention to physics, especially since Born had taken direct aim at wave mechanics with a second paper on atomic collisions. Born's June paper on atomic collisions had made good use of Schrödinger's wave equation and de Broglie's matter waves to, ironically, demonstrate the existence of quantum jumps. Born was unabashedly enthusiastic about wave mechanics, referring to it in his paper as "the most profound formulation of the quantum laws."[19] But by the time Born's second collision paper came out in July, it was obvious that he had switched sides and had joined the rest of the Copenhagen and Göttingen crews in their collective opposition to wave mechanics.

The aim of every decent theory, according to Born, was to advance Bohr's concepts of stationary states and quantum jumps.[20] Born therefore reinterpreted the square of Schrödinger's wave function so that it determined the probability of an atomic state. It did not represent anything visualizable or real; it was just a number, a probability. But this was not the probability that Boltzmann, Planck and Einstein had used to figure out what very large numbers of molecules, imaginary resonators or light quanta were doing, which, under Newtonian physics, they *could* know—at least in theory. This was new. The old probabilities, said Born, were based on the inability of physicists to know what each particle was doing, and determined instead what they were most likely doing. His new probability, however, was based on the inherent unknowability of random events in the quantum world.

Born got the idea from Einstein. In their long discussions, Einstein had mused about light quanta as being guided by a "ghost field" of waves, with the waves determining where the light particles went. Born had reinterpreted Schrödinger's wave function to be only the probability that the "ghost field" would randomly toss up a particle in a particular state, and in doing so he had turned Schrödinger's ψ into yet another mathematical symbol without any physical meaning.

Einstein was not exactly thrilled that Born had given him credit for the "ghost field" idea in his published paper. Einstein had never published the idea himself because he didn't agree with it. Born had been quite hurt by Einstein's rejection of matrix mechanics, and now the two men had arrived at irreconcilable differences. Born was fine with a random and probabilistic fundamental reality. Einstein was vehemently opposed, and there was nothing Born could say to convince him otherwise.

Heisenberg, meanwhile, was comfortably ensconced in his new quarters at the Bohr institute. The top floor of the institute had been turned into guest quarters for assistants and visiting researchers, and when Heisenberg arrived in the spring of 1926 he was the first to take up residence there. The new house for the Bohr family on one side of the institute building and the lab on the other were finished, and Niels, Margrethe and their five boys had finally settled into their new home.

Bohr may have been thinking a lot about quantum physics, but he hadn't published anything particularly meaningful since the failed BKS paper in 1924. The sudden appearance of a completely different theory of the quantum at the beginning of 1926 had rattled him. He had difficulty wrestling with the implications of matrix mechanics, but at least it seemed to signify that finally a sensible solution might be found to the riddle of the quantum. Then Schrödinger had dropped his little bombshell of wave mechanics.

Bohr had called a colloquium as soon as the new wave mechanics paper arrived. Pauli was still at Copenhagen at the time, and had been on the lookout for it. "Here," Bohr told the assembled researchers, "is an exceedingly powerful and fertile method."[21] The new theory excited considerable debate at the institute, which continued in the hallways and through mealtimes. What did it mean to have two such different theories? It wasn't long before they were all feeling the strain, so it was something of a relief when Pauli could finally tell the Copenhagen scientists that the two theories were equivalent to each other—just different means of arriving at the same solutions. About the same time, Ernest Rutherford invited Bohr to Cambridge, and he was glad to go. "I look forward to private discussions about our present theoretical troubles," Bohr told Rutherford, "which are of an alarming character indeed."[22]

Bohr's energy was already drained by his never-ending administrative and fundraising duties and by the problems he'd had to deal with to finally get the Bohr house and new lab finished—behind schedule, of course. And it was frustrating that, once again, the hope of finding a way out of the great quantum muddle had been thwarted by surprising new developments. Figuring out the quantum was like trying to nail jelly to the wall. By the summertime

Bohr was once again in a state of exhaustion. He succumbed to a nasty bout of the flu and was out of commission for much of the summer.

Bohr was conscientious about taking care of scientists who stayed at Copenhagen, especially the younger men, who often had little money. He endeared himself to many a needy young physicist by finding a place for him to sleep or quietly slipping him a little money for food. However weary and frustrated Bohr might have been, he still found time before he became ill to put together some money for Jordan to go to Vienna for treatment for his speech impediment. Jordan sent a note to Bohr at the end of July, outlining his plans.

> I expect to use the now starting vacation months for restoring my health. On the advice of my doctor, I shall go first to Bad Pyrmont for several weeks. Further, I intend to consult a neurologist, A. Adler in Vienna . . . At the beginning of the winter semester I plan to return to Göttingen and expect, if the treatment has led to the desired success, to obtain [my] *Habilitation* there.[23]

The mineral springs and luxury spas at Bad Pyrmont were a popular destination for people taking "the cure" or just seeking respite from the demands of the world. It's not clear if Jordan was suffering from a particular illness in the summer of 1926 or whether his nerves were getting the better of him, but one of Pauli's colleagues from Hamburg had highly recommended a visit to Alfred Adler, a renowned psychologist, where Pauli's friend had been cured of "nervous gastropathy."[24] Jordan may indeed have taken "the cure" in Bad Pyrmont, but there is no evidence he sought out treatment for his stuttering. Or if he did, it had no apparent effect.

Heisenberg and Born had made it clear they did not share Jordan's initial enthusiasm for wave mechanics, when, as soon as wave mechanics came out, he'd been able to see the potential for describing matter by quantizing Schrödinger's wave function ψ. Jordan had his doctoral degree, but he still had to consider his habilitation. It wasn't long before he became even more vociferously critical of wave mechanics than Heisenberg.

In the summer of 1926, Ehrenfest and Einstein had taken a time out from the heated debate amongst physicists of the pros and cons of wave mechanics versus matrix mechanics. They both had serious reservations about the direction quantum physics was taking, but stepped out of the fray into another place altogether—the fifth dimension. Einstein had good reason to be interested in higher dimensions, since his theory of general relativity permitted the use of any number of dimensions. And when higher dimensions were involved, strange things happened to the classical physics of four dimensions (three space and one time).

In 1917, Ehrenfest had invited Finnish physicist Gunnar Nordström to Leiden. Nordström had just published a paper on five-dimensionality, and Ehrenfest was curious to know more. Tatyana Ehrenfest was a mathematician with a special interest in geometry, and she too found higher dimensionality most intriguing. Nordström's visit spurred Ehrenfest to revisit the question that had been posed by philosophers from Aristotle to Kant: why are there three dimensions in the observed universe instead of some other number? Ehrenfest's conclusion was simple: planetary orbits are only possible in three space dimensions, otherwise planets would spiral out of control; waves move differently in other dimensions and therefore it would be impossible to transmit information; and quantization of Bohr's atom could exist only in three-dimensional space. In four space dimensions there would be an infinite number of allowed orbits, no lowest energy state (the orbit closest to the nucleus) and no spectral lines.[25] In other words, Bohr's quantized atom in three space dimensions transformed into a classical atom in four.

Ehrenfest's paper was published in 1917 in the Dutch *Proceedings of the Royal Society*, where it got about as much attention as his 1911 review of the issues in quantum physics prior to the First Solvay Conference. It was ignored. There was a war going on. Undeterred, Ehrenfest kept up his interest in higher dimensions, and in the summer of 1926, when he heard about the five-dimensional work of the Swedish physicist Oskar Klein, he invited Klein to Leiden. George Uhlenbeck, Ehrenfest's assistant, was also quite familiar with higher dimensions, since just about any theory he and Ehrenfest were working on would also explore what might happen in higher dimensions.

Klein had been considering for some time that the fifth dimension of space-time might have something to do with quantization, and it was exciting to consider the possibility of unifying the quantum and classical worlds. Einstein was also working on such a theory. For years he too had been trying to figure out how to reconcile Maxwell's electromagnetic equations with his own equations for gravity. "When Oskar Klein told of his ideas," said Uhlenbeck, "which would not only unify the Maxwell with the Einstein equations but also bring in the quantum theory, I felt a kind of ecstasy. Now one understands the world!"[26]

Uhlenbeck's delight was, of course, a little premature, but understandable given the circumstances. Anything seemed possible in the summer of 1926, with two versions of quantum mechanics to choose from, and the fifth dimension offering a link between the classical and quantum worlds.

Einstein almost believed it too. The idea of a fifth dimension in general relativity had first been proposed in 1919, with recent refinements by Klein, but it was really more of a mathematical construct. There was, after all, no actual evidence of a fourth physical dimension. But the concept had a more serious flaw: when it came to calculating such well-known values as the mass and charge of an electron, the proposal was way off base. The excitement over five dimensions quickly dissipated.

Einstein and Ehrenfest had not stopped following the wave versus matrix conflict; the fifth dimension was just a diversion. Einstein spent time with both Heisenberg and Schrödinger when they each lectured in Berlin on their respective versions of quantum mechanics, but he was turning more and more of his attention to the puzzle of how to link the geometry of gravitational theory with the algebra of quantum physics, which effectively put him on a different path from that of most other physicists.

Nonetheless, Einstein was once again staying with Ehrenfest in Leiden in the fall of 1926, where they were attempting to work out Dirac's latest mathematical/physics effort. It wasn't easy, because Dirac had a particular fondness for making up his own mathematical notation. He still worked alone in his college room, and even when he was home in Bristol he spent most of his time working in his bedroom, emerging only to use the bathroom or to get something to eat.[27]

In the spring of 1926, Dirac had completed his doctoral degree, and he'd invented q-numbers to represent non-commutating quantities and c-numbers to represent classical numbers (which always commute). He'd developed his own version of matrix mechanics using q-numbers to represent matrices, and then had written a version of q-number algebra that did not contain any reference to physics at all, something that appealed only to other mathematicians. Jordan thought it "very beautiful,"[28] but most physicists, including Einstein and Ehrenfest, were not quite so eager to exchange physics for the abstract formalism of mathematics, and Dirac's work was not easy for them to follow.

In August, Dirac had published a paper on the relativistic wave equation that Schrödinger had abandoned in December 1925. The main reason his version hadn't worked was that he hadn't known about spin then, and a relativistic equation had to include spin to produce the correct results. The early fall produced a flurry of papers by physicists who each had his own take on the relativistic wave equation. But it was Dirac's paper that Ehrenfest and Einstein were trying to figure out. "Einstein is currently in Leiden," Ehrenfest wrote to Dirac at the beginning of October. "In the few days we have left, he, Uhlenbeck and I are struggling for hours at a time studying your work, for Einstein is eager to understand it. But we are hitting a few difficulties, which—because your presentation is so short—we seem absolutely unable to overcome."[29]

Born was a little overcome himself in the fall of 1926. Göttingen was fast becoming a victim of its own success, and in September the university was inundated by students and visiting researchers. The little college town turned into a bit of a madhouse. One of Born's students considered it a wild and crazy time, with "people who were highly bizarre, genially mad, unworldly, and completely, decidedly difficult in their behaviour."[30] On the positive side, there were parties and noisy street parades, but striking an uglier note were a number of ardent National Socialist students who had quietly begun compiling a list of Göttingen's Jewish professors.

With Heisenberg in Copenhagen and Jordan working on his doctoral thesis, Born was feeling the strain of trying to manage the madhouse. His relationship with his students suffered as he became even more distant and difficult. Walter Elsasser, one of Born's doctoral students, found himself a victim of the "mill." When Born assigned Elsasser an uncomplicated thesis and then ignored him, it seemed as if Born wanted to get rid of him as quickly as possible. "He did not show the slightest interest in me as a scientist or a person," said Elsasser, who admitted he wasn't an exceptional student. "I seemed to be just another one of the nuisances that students sometimes represent for their professors."[31]

Fortunately, an American doctoral student, Robert Oppenheimer, had arrived at Göttingen to work with Born, and he helped Elsasser prepare his thesis before heading off to Leiden to study with Ehrenfest.

Wave mechanics seemed to have an inspiring effect on physicists, who were applying Schrödinger's methods to all manner of physics problems. Even Heisenberg resorted to using wave mechanics to successfully approximate the spectrum of the helium atom, something he hadn't been able to accomplish with matrix mechanics. It was quite apparent that Schrödinger's wave mechanics had won the day, and there was little the Göttingen and Copenhagen schools could do about it.

But there was still a window of opportunity to salvage matrix mechanics. Schrödinger's interpretation of wave mechanics was based on continuity, which did away with the need for quantum jumps, while matrix mechanics was based on discontinuity, with quantum jumps. Perhaps it was still possible to convince the physics community that, while Schrödinger's mathematics was undeniably a useful tool, his interpretation was all wrong and the Göttingen and Copenhagen one was right.

Schrödinger headed off to Copenhagen at the end of October to give a series of lectures on wave mechanics. Bohr was keen to talk to Schrödinger, and despite the fact that both men had been involved in theoretical physics in Europe for nearly two decades, they would be meeting for the first time.

Chapter Fourteen

DARK NIGHT OF
THE SCIENTIFIC SOUL

*By far the most usual way of handling phenomena so
novel that they would make for serious rearrangement
of our preconceptions is to ignore them altogether, or to
abuse those who bear witness for them.*

WILLIAM JAMES

Aside from their interest in quantum physics, Erwin Schrödinger and Niels Bohr had little in common when they met in the fall of 1926. Schrödinger had grown up as an indulged and spoiled only child, and turned into a self-centred man. He was a sensualist with a fine appreciation of food, wine and theatre, and he expected his wife to devote herself to his care and comfort, and to willingly accommodate his affairs. But he had no children. It appeared that Anny was unable to bear children, and Erwin regretted not having a son.

Bohr, on the other hand, had five lively young boys, and he made certain he had time to play games with them and talk about their concerns when they were together at the dinner table. Bohr also had a wife whom he simply adored. Bohr was a father figure to the institute's students and visitors, and he rarely worked alone. He needed someone with whom he could talk out his ideas in order to help clarify his convoluted thinking, and he needed someone to write down his ideas in the form of dictation. Bohr had built a school around himself, filled with bright young people and visiting scientists, all talking and calculating, engaged in a never-ending scientific dialogue. Bohr had a ready pool of assistants to draw from. Schrödinger worked alone.

Bohr intended Schrödinger's visit at the beginning of October to be a productive one. He planned a full-scale debate on everything they knew—and didn't know—in atomic physics. News of the Schrödinger lectures and seminars attracted a large number of scientists, as well as those already in residence at the institute. The discussion was lively, with lots of debate and equations scribbled on the blackboards. But it got even livelier when Schrödinger started arguing that quantum jumps should be abandoned since they were readily replaced by waves vibrating around the nucleus of the atom that simply shifted from one frequency to another. Biographer Ruth Moore described the reaction: "Half a dozen physicists were shouting objections and questions. Bohr, his pipe forgotten, was pacing the room. Everyone was haranguing his neighbor."[1]

Bohr finally broke through the uproar to restore order, but the debate over quantum jumps continued in earnest for the rest of the week. Bohr was sometimes oblivious to those around him who might want to sleep, keeping discussions going until the wee small hours of the morning. Since Schrödinger was staying with the Bohr family, he had no escape from Bohr's relentless questioning. According to Werner Heisenberg, who was a witness to the debate between Bohr and Schrödinger, "It will hardly be possible to convey the intensity of passion with which the discussions were conducted on both sides, or the deep-rooted convictions which one could perceive equally with Bohr and with Schrödinger in every spoken sentence . . . So the discussion continued for many hours through the day and night without a consensus reached."[2] Schrödinger refused to accept quantum jumps or Born's probabilistic interpretation. He wanted a classical interpretation of the behaviour of an atom, and Bohr would not budge on discontinuity and stationary states.

Before the week was over, Schrödinger was ill and resting in bed. That didn't stop Bohr, however, who simply sat on the side of the bed and kept arguing. Bohr very nearly succeeding in persuading him to accept quantum jumps, said Schrödinger afterwards, but he held his ground, although only just.[3] It took a lot of stamina to withstand Bohr's dogged determination, which meant that he sometimes simply exhausted his opponents into surrender.

Whatever else Schrödinger accomplished in Copenhagen, he had inspired Bohr to get back to work on developing a theoretical explanation of quantum physics. "We had great pleasure from the visit of Schrödinger," Bohr told Ralph Fowler. "After the discussions with [him] it is very much on my mind to complete a paper dealing with the general properties of the quantum theory."[4]

Schrödinger returned to Zurich, his confidence shaken by the gruelling debates at Copenhagen, and wondering if he hadn't taken a wrong turn somewhere with wave mechanics. But he rallied, and in a letter to Bohr after the trip he reiterated his belief in the importance of creating representations or pictures that would make the world intelligible, "even if a hundred attempts miscarry."[5]

Schrödinger was also taking on Max Born, challenging his insistence that quantum jumps were an integral feature of quantum mechanics. It wasn't necessary, Schrödinger wrote at the beginning of November, to dogmatically insist that discontinuities be part of the theory, especially when he had already demonstrated that the transition from one atomic state to another could be correctly calculated using continuous wave theory. Schrödinger pointed out that Born and his colleagues appeared to have become "addicted" to the old ideas of quantum jumps and stationary states. Born replied promptly, admitting that it probably was a good idea to explore all possibilities, but adding that he wasn't going to because he had a feeling that one couldn't dispense with quantum jumps, and he was going to go with that feeling.[6]

Whatever the debate between the creators of the wave and matrix models, physicists were voting with their pens and pencils. Scientific journals were deluged with papers using Schrödinger's wave equations and ignoring matrix mechanics, simply because the familiar wave equations could be successfully applied to a wide range of physics problems where matrix mechanics was of no use.[7]

Paul Dirac arrived at the Bohr institute in the fall, just about the same time as Schrödinger. Dirac, however, had no desire to involve himself in the philosophical debates that raged through the offices and hallways. He didn't talk; he did equations. Even when he gave a lecture, he tended to carefully write his equations on the blackboard, perhaps offer a statement or two, and that was it.

Dirac's visit to Copenhagen gave him the opportunity to meet Heisenberg, Pauli and others who were working on quantum physics, but he still maintained his isolation. "I learned to become closely acquainted with Bohr, and we had long talks together, long talks on which Bohr did practically all the talking."[8]

A young scientist, when tapped by Bohr to assist him with his dictation, had to put his own work aside and devote his time exclusively to Bohr. The designated "victim" had to sit and take notes, and was not allowed to pace; only Bohr could pace. One of Bohr's assistants described the process:

> When Bohr decided he wanted to work on a paper the victim was called into his office . . . There he stayed throughout the day while Bohr roamed around the room circling the victim's desk every few minutes . . . When the victim got over the feeling of being used, he realized that he had been given a glorious opportunity. He was witness to the mind of Niels Bohr in action.[9]

Dirac was not one to get caught up in the hero-worship of Bohr that had become so much a part of the "Copenhagen spirit." Besides, Bohr expressed himself with words while Dirac expressed himself with equations, and since Bohr didn't do equations, Dirac's stint as

Bohr's "victim" lasted all of half an hour.[10] Dirac had research he wanted to work on anyway, and Bohr's assistants usually had little time to work on their own ideas.

As Bohr's right-hand man, Heisenberg had little time to himself, but ostensibly they were working together on the most pressing issue of the day—producing a paper on the "general properties of the quantum theory." It might have helped if the two men had first sat down and discussed what they were each trying to accomplish. It might have saved them a lot of grief, because they were working at cross-purposes right from the start.

Heisenberg's objective was simple: he wanted to prove the superiority of matrix mechanics over wave mechanics, and demonstrate the particle nature of the world. This was, of course, in direct opposition to Schrödinger, who wanted to replace all particles with wave packets and demonstrate the wave nature of reality. But Schrödinger did not have a doubting father peering anxiously over his shoulder, or the worry that he'd scuttled his career advancement by turning down an important job offer. Schrödinger advanced his interpretation of the atom on deeply held philosophical beliefs; Heisenberg was finding he could be quite flexible, philosophically speaking.

Bohr, on the other hand, wanted to build a theory that incorporated both waves and particles, because he considered wave-particle duality to be one of the key characteristics of the quantum world. It wasn't going to be an easy task. Bohr was also very attached to stationary states (allowed orbits). They had been part of his understanding of the quantized atom since 1913, and he wanted to keep them. Schrödinger's interpretation did away with quantum jumps, so Bohr had to come up with some way to keep the waves *and* the quantum jumps.

Since Heisenberg and Bohr had different goals, it's hardly surprising that they were continually butting heads. Heisenberg had learned to stand up to Bohr, and he was determined to develop a theory to suit his purposes. So was Bohr.

The two men had taken it upon themselves to develop a new quantum theory, even if they were at odds about what the interpretation should be. Schrödinger was leaving the field clear by taking himself off to the United States before Christmas to lecture on wave mechanics. Dirac had no interest in developing an interpretation, and Pauli was holding himself aloof because he had his own reservations about both matrix and wave mechanics. That left just Bohr and Heisenberg working on a theoretical model. They didn't invite Born to join them.

Born learned in the fall that he had made the short list to replace the retiring Max Planck in the chair of theoretical physics in Berlin, along with Sommerfeld and Schrödinger. Given that there was little likelihood of Sommerfeld leaving Munich, the competition was really between Born and Schrödinger.

Born had had his eye on replacing Planck since the earlier time in his career when Einstein had advised him to take the professorship in Frankfurt rather than stay in Berlin in the hope of succeeding Planck. It's not clear how badly Born wanted the job, but he let Planck know that, if he were called to take the Berlin position, it might be difficult for him because James Franck had turned down a position in Berlin two years earlier to stay in Göttingen with Born; so it would be disloyal for Born to now abandon Franck. As well, the Nobel committee had just announced that Franck and a former research partner were to be the recipients of the delayed 1925 prize for physics. Perhaps it suited Born better to continue his collaboration with his friend and Nobel laureate than to move to Berlin.

The Nobel Prize for Franck and his colleague was a shot in the arm for the Göttingen–Copenhagen version of quantum mechanics. In 1914 the two men had been doing experiments using mercury atoms. Bohr later persuaded them that the results of their experiments demonstrated the quantized energy levels of Bohr's stationary states. It wasn't actually what Franck and his colleague had intended to do, but that's what they got the Nobel for.[11]

Göttingen's December celebration of Franck's shared Nobel Prize was one of the few bright spots in Born's otherwise distressing life. Heisenberg had written a paper in November using Born's probability interpretation for Schrödinger's squared wave function without bothering to cite Born's paper. In fact, Heisenberg hadn't mentioned him at all, and Born was deeply offended. Born and Einstein weren't talking anymore, either; their exchange of letters had stopped. And Hedi was more remote than ever, so it perhaps helped ease Max's loneliness to have the charming Maria Göppert to talk to. Maria had started university at Göttingen in 1924, studying mathematics, but Max thought she would do better working with him and persuaded her to switch to theoretical physics. The Göpperts were a well-known academic family in Göttingen, and Maria had always expected to be a professor or researcher herself. She'd grown up next door to David Hilbert, watching him lovingly tend his rose garden with a portable blackboard always at hand in case an equation came to mind, and the Göpperts were good friends of the Born and Franck families. Maria was much younger than Max—twenty to his forty-five—but she was a very good listener, and Max needed someone to talk to.

Schrödinger had his own female friends to talk to. He did not go to Arosa for Christmas 1926. Instead, he spent the holiday with the twins Withi and Ithi and their mother at a ski lodge. Erwin was romancing fifteen-year-old Ithi, but it was more of a wooing than a serious attempt to get her into bed with him. However, the holiday didn't go well, and his ardour was dampened by the lack of heat in their rooms and a sprained ankle.[12]

The end of 1926 was hardly the erotic and creative venture of Schrödinger's previous year. When Erwin and Anny boarded the ship for the United States at the end of December, it turned out that Anny was not a good sailor, and she spent most of the trip across the Atlantic in their cabin being sick. Erwin did not care for the "repulsive" class of passengers on the ship, so he was left with the unpleasant choice of spending time on deck with distasteful companions in the chill of a January ocean crossing or listening to Anny's retching. It didn't help Erwin's mood when they finally arrived at New York City and he couldn't get a drink because of Prohibition.[13]

With Schrödinger off to America, there was no one in Europe to advocate the wave theory, whereas the Copenhagen and Göttingen schools were enthusiastic and sometimes militant in their promotion of their evolving version of quantum mechanics. Sommerfeld, as the head of the Munich school, could have thrown his weight behind Schrödinger's visualizable, more-classical-than-not theory, but he had reservations. Schrödinger had yet to explain the photoelectric effect via waves, nor had he resolved the problem with spreading out of the tiny wave packets that he wanted to use to represent electrons.

It was getting harder and harder to tell who was on whose side in the development of quantum physics. It was a messy business. The new theories were in constant flux, so what might look good to one of the quantum creators on Monday could have evolved into something quite offensive a week later. Because the theories were confusing and incomplete, it was quite possible to hold an irrational or illogical set of beliefs, especially if one's personal beliefs did not match a publicly stated position on an issue. Personal ambitions, strained friendships and powerful conflicting agendas muddied the waters and made for stormy times indeed.

In Copenhagen, Bohr and Heisenberg started out the new year in full battle mode, but they weren't fighting Schrödinger anymore. They were battling each other.

Pauli had not involved himself directly in either wave or matrix mechanics. In 1926 he had finally completed the comprehensive article on quantum physics—another encyclopedia-type work—which had occupied much of his time during the previous year. His article dealt with what was understood about the quantum world before the development of matrix and wave mechanics, a review of the bewildering experiments and the difficult time physicists were having trying to make sense of the experimental results when everyone was groping in the dark.

Pauli was still a lecturer in Hamburg, and after four years he hadn't been offered a professorial position. He wasn't a particularly good lecturer since he had little patience with students who couldn't keep up with his thinking, and he resisted simplifying topics to make them easier to understand.[14] And perhaps his sharp tongue had not endeared him to potential employers.

Heisenberg was keeping an ear to the ground for a professorial position for himself, but nothing had come up since he'd turned down the Leipzig offer. He hadn't yet succeeded in giving matrix mechanics the necessary facelift to make it more appealing to scientists, so Heisenberg decided that if physicists found Schrödinger's theory more acceptable because it was visualizable, then he would make matrix mechanics visualizable as well. It was Pauli who provided the inspiration.

Even though Pauli had absorbed positivist philosophy "with his mother's milk," so to speak, he was not an ardent positivist. Despite the fact that matrix mechanics had been initially constructed on the principle of using only observable entities—a fundamental tenet of positivism—Pauli did not see why it was necessary to assume that electrons in an atom were unobservable on principle.[15] In the fall of 1926, Pauli wrote to Heisenberg that he believed, "with all the fervor of my heart,"[16] that the elements of the matrices corresponded with the details of actual particles in the atom, and therefore could be considered visualizable in principle, if not in actual practice. There was, after all, still no way to actually "see" the inner workings of the atom.

Heisenberg was very enthusiastic about Pauli's letter, which quickly made the rounds at Copenhagen. It was just what Heisenberg needed to develop his own theory of quantum mechanics. Pauli's view supported visualizability, the particle nature of the quantum world, and the return of the electron and electron orbits. This was, of course, a decided reversal of the thinking that had led to Heisenberg's 1925 observables-only paper, where the inner workings of the atom were unknowable because they were unobservable. But his primary goal in developing a new theory was to convince physicists that matrix mechanics was right and wave mechanics was wrong. If that required some interesting philosophical gymnastics, so be it. It wasn't as if Heisenberg had any strong philosophical beliefs of his own.

At the end of November, Heisenberg's viewpoint got a boost from Dirac, who had figured out how to transform Schrödinger's wave function into a matrix that contained all the physically meaningful information that was known in quantum mechanics. Jordan had come up with a similar transformation theory at about the same time. Thus, Schrödinger's wave mechanics could be transformed into matrix mechanics, with the wave equation reduced to little more than helpful mathematics for particle matrices.

Bohr, meanwhile, was struggling to develop an interpretation that would validate his long-held belief in both stationary states (allowed orbits) and quantum jumps. He wanted to reconcile his beliefs with wave theory, but he was making little progress.[17] With Heisenberg working feverishly on a theory that validated only the particle nature of the quantum world via matrix mechanics, there was a lot of tension in the air at the institute in Copenhagen. Their interpretive positions clashed mightily, but neither was willing to back down on his preferred version of the theory. Scientist and author David Lindley described the atmosphere:

> In Copenhagen, the two men would spend hours together during the day, Bohr talking always in his unrelenting, insistent way while Heisenberg, animated and agitated, struggled to interrupt. In the evenings they would often continue the haggling as they took a turn around the pleasant grassy park that adjoined the institute. Often, too, even late into the night, Bohr

would knock on the door of the attic room at the institute where Heisenberg was staying, offering just a small clarification or emendation of what he had been trying to say earlier. Not infrequently those footnotes to the day's discussion would sprawl into the small hours. Bohr would stick to no fixed schedule. Whatever had to be said had to be said there and then.[18]

The two men had planned to go on a skiing trip to Norway in February, but Bohr ended up going by himself. They needed to get away from each other, so the break was a good idea. Relieved of Bohr's dominating presence, Heisenberg had time to think. And Bohr needed a rest as well. Oskar Klein, who was staying at Copenhagen at the time, said Bohr was very tired, and that the new quantum mechanics "caused him both much pleasure and very great tension."[19]

Heisenberg, freed from Bohr's unrelenting attentions, sat down and wrote a very long letter to Pauli, explaining in detail the new theory he was working out. The position of a particle would be observable in principle; the path of a particle would be discontinuous and therefore only describable statistically; and hence the position and momentum of a particle would not be able to be determined at the same time.[20]

None of these ideas were news to Pauli. Dirac had developed his transformation theory the previous fall, but it had a bit of a kink in it, in that the portrayals of the position p and momentum q didn't seem to fit together properly. Pauli noticed it too, and was puzzled by the fact that if one knew the momentum of a particle, one could not know its position. He had discussed that problem in a letter to Heisenberg back in October. "You can look at the world with the p-eye," said Pauli, "and you can look at the world with a q-eye, but if you want to open both eyes at the same time, you will go crazy."[21]

Pauli had also discussed the idea that electrons could be considered observable in principle. Discontinuity had long been debated as a fundamental characteristic of the quantum world, except, of course, in Schrödinger's theory. By allowing that the path of a particle was also quantized—that an electron could be imagined to appear only at discontinuous points along a path—it made sense to Heisenberg that the position (the point on the path) and momentum (the travel between the points) could not be determined at the same time.

Heisenberg wasn't the only scientist who was thinking about uncertainty in measurement. A Dutch experimentalist and a Swedish experimentalist had both published separate papers in 1926 about the limits on the accuracy of measurements due to the constant zigzagging of particles (Brownian motion). The paper by the Dutch experimentalist was published in the same issue of the German journal in which Jordan's version of the transformation theory appeared, so it is very likely that Heisenberg would have seen that article when he read Jordan's.[22]

Born had also weighed in on the limits of accuracy. In his earlier atomic collision paper, Born had specifically addressed the fact that quantization meant units in the quantum world were always limited to values multiplied by whole numbers—Heisenberg's half quanta having long since been abandoned. Since there could be no fractions of a unit or partial unit, physicists had arrived at the bottom of nature's scale with quantum physics.[23] Heisenberg expanded the idea of limits to nature's scale by making Planck's constant h the limit to what could be known simultaneously about changes in position and momentum, and they could never be known with certainty. This was a turning point for Heisenberg's theory, because it led him to pursue a visualizable interpretation of quantum mechanics through thought experiments based on the limits of measurement. Heisenberg wrote out all his ideas in a letter to Pauli at the end of February, in an attempt, he said, to "get some sense of his own considerations"[24] as he groped towards a consistent theory.

According to historian of science Mara Beller, Born was motivated to develop a theory of quantum physics because he considered quantum mechanics to be nothing more than a tool for describing and predicting experimental results: but not Bohr or Heisenberg.

> Most physicists, Bohr and Heisenberg included, wanted more: some feeling of understanding, of illuminating, or explaining the kind of world that quantum formalism describes. The need for this kind of metaphysical grasp is not merely psychological but social as well—the power of a successful explanation and the power of the effective legitimation and dissemination of a theory are connected.[25]

Both Bohr and Heisenberg were aware that coming up with the first comprehensive and complete theory of quantum physics would allow them to put their particular stamp on it. Of course, Heisenberg was also motivated by the need to prove "his" matrix mechanics superior to Schrödinger's wave mechanics. And he'd still had no job offers.

Heisenberg ended his letter to Pauli with an apology, wondering if any of his deliberations were of any value. Pauli did, apparently, question the uncertainty relation Heisenberg proposed, since it didn't arise out of the arguments of the paper and such an interpretation did not appear to be self-evident. In his response to Pauli at the beginning of March, Heisenberg admitted that his uncertainty interpretation was somewhat dubious and told Pauli that he might rewrite it or drop it altogether.[26] He did neither.

With Dirac's claim that his transformation theory provided all the information needed to make experimental predictions, Heisenberg was satisfied that matrix mechanics was completed, and he had his own theory to go with it—this, despite Jordan having just published a paper arguing that quantum physics, as it stood at the beginning of 1927, was in fact incomplete. Belief in the finality of a theory, Beller has noted, is an emotional choice or ideological stand, not a scientific judgment. Heisenberg believed that he had created a final and complete theory of quantum physics, and anything that didn't fit, such as Schrödinger's continuity, was obviously just wrong.

In short order, Heisenberg revised his February letter to Pauli and turned it into a journal article, which he submitted for publication before showing it to Bohr when he returned from his ski trip to Norway around March 18. By this time Bohr had decided that the best way to allow that a particle could be both a wave and a particle at the same time, with the experimental apparatus determining which side of its dual nature appeared, was to do away with any kind of space-time description of the quantum world. This paradox could be satisfactorily dealt with if one declared the quantum world non-visualizable. But Heisenberg's paper was based on the premise that the quantum world *was* visualizable. Oh, dear.

Chapter Fifteen

SOLVAY PRELUDE

When people are fanatically dedicated to political or religious
faiths or any other kind of dogmas or goals, it's always because
these dogmas or goals are in doubt.

ROBERT J. PIRSIG,
philosopher

The provisional invitations to the Fifth Solvay Conference were
mailed out at the end of January 1927. The tentative conference
agenda had been set the previous April by the Solvay scientific committee, which included Hendrik Lorentz, Albert Einstein, Marie
Curie and Paul Langevin, among others. The topic chosen was
"The quantum theory and the classical theories of radiation." The
agenda listed seven reports: Lawrence Bragg, Arthur Compton
and Charles Wilson would each report on their experiments; Louis
de Broglie would report on "interference and light quanta"; Hans
Kramers on the BKS proposal; Einstein on applications of statistics
to quanta; and Werner Heisenberg on the "adaptation of the foundations of dynamics to the quantum theory," with Schrödinger
listed as a possible substitute for Heisenberg.[1]

Einstein wasn't particularly keen to give a talk, so he recommended that he be replaced by Schrödinger. Lorentz didn't immediately agree because, at the beginning of the year, he still wasn't
quite sure about Schrödinger's wave mechanics. But Lorentz
changed his mind when, by happenstance, both he and Schrödinger

ended up in Pasadena, California, in March. Schrödinger was wrapping up his west coast visit and then heading across country to give a few more lectures before returning home. Lorentz was lecturing at the California Institute of Technology until the end of March. They got together a number of times in Pasadena, and Lorentz took the opportunity to sound out Schrödinger about giving a report on wave mechanics at the Solvay conference.[2]

Giving a report at a Solvay conference was not a task to be undertaken lightly. The point of the conferences was to thoroughly investigate a handful of the newest or most significant ideas in physics. To allow enough time for the reports and the discussion to follow, one report was scheduled for the morning session and a second for the afternoon. That meant the scientist giving a presentation not only had to prepare a detailed and comprehensive report on the selected topic, but he (or she, if Curie was reporting) had to field questions for the entire session from some of the highest-calibre physicists in the world. It was not for the faint of heart.

The agenda for the October conference was still evolving in the spring, which was just as well since quantum physics was evolving too. In January, American experimentalist Clinton Davisson and his partner published a short article in the journal *Nature* confirming the existence of de Broglie's matter waves. The Americans had done the experiments several years earlier after an equipment malfunction had produced some strange results. These were the results that Walter Elsasser had suggested might be evidence of matter waves. Davisson had no idea of the significance of the 1923 experiments until he was listening to a talk being given by Max Born at a 1926 Oxford meeting, in which Born cited Davisson's experiments as confirmation of electrons as waves. Born, James Franck and a few others met with Davisson after the meeting to give the surprised experimentalist the details of his inadvertent "discovery."[3]

Louis de Broglie was no doubt pleased to see the published confirmation of his matter waves, followed not long after with a second verification by George Thomson (J.J.'s son) in England. But in early

1927, de Broglie was deeply immersed in developing a completely new theory of quantum mechanics. He did not share Schrödinger's desire to make the quantum world all about waves, nor did he share the Göttingen wish to make it all about particles. De Broglie's view was more like Bohr's, in that he too believed in a quantum world of both waves and particles. The challenge he faced, just as Bohr did, was to reconcile the paradox of simultaneous—yet physically impossible—wave-particle duality. Past conflicts meant there was little love lost between de Broglie and the Copenhagen school, so neither of them knew what the other was doing. De Broglie worked alone. He was lecturing at the Sorbonne and living with his mother in the family mansion. Louis had his brother Maurice to talk to, of course, but Maurice was an experimentalist, not a theorist.

In his 1924 thesis, de Broglie had been clear that he wanted to reconcile the wave and particle natures of light. "When two theories, based on ideas that seem entirely different, account for the same experimental fact with equal elegance, one can always wonder if the opposition between the two points of view is truly real and is not due solely to the inadequacy of our efforts at synthesis."4

In early 1927, de Broglie was working on a wave-particle synthesis when he decided something was going to have to go. Ordinary classical physics wasn't going to do the trick. Bohr had been prepared to sacrifice conservation of energy and momentum in the BKS paper, whereas de Broglie was now willing to give up classical conservation of momentum in his theory. He proposed that all particles in motion were accompanied by waves that guided them, and therefore the trajectory of a particle could appear to change without cause when its path was being altered by its companion wave. This would appear to be a violation of conservation of momentum—as well as Newton's first law of inertia—unless one allowed that the hidden guide wave caused the particle to change course.

De Broglie published a short paper in a French journal at the end of January, and kept working to expand his theory. He developed what he called a "double solution" equation which produced, he believed, one solution that represented a particle and another that represented a wave. Thus, in his theory, he had reconciled waves and particles via the double solution. He also argued that the "pilot wave" which guided a particle was the same as Schrödinger's wave function ψ, which also represented Born's statistical probability in wave and matrix mechanics; but, unlike Born's approach, de Broglie's waves were real, and, unlike Schrödinger's wave model, so were his particles.

De Broglie published his deterministic, visualizable theory in *Journal de Physique* in May. The paper, "Wave mechanics and the atomic structure of matter," would be the basis for his talk at the Solvay conference that fall.[5]

Meanwhile, back in Copenhagen, Bohr's attempt to reconcile wave-particle duality was not going very well.

Bohr wanted the interpretation he and Heisenberg were working on to include waves. He liked the idea of an electron as a tiny wave packet with one or all allowed energy states present at the same time. Once the electron jumped to a specific stationary state (allowed orbit), only the vibration that characterized that particular state remained, but how such a thing could happen was not visualizable. To incorporate stationary states—which he had no intention of giving up to either wave or matrix mechanics—with Schrödinger's wave, Bohr rendered the workings of an atom non-visualizable by giving up any possible space-time description of the electron.

None of this sat well with Heisenberg, and the language of their debates was becoming harsher. After Bohr had returned from his ski trip in March, he and Heisenberg had got into a huge row over Heisenberg's uncertainty paper. Aside from the fact that they were working at cross-purposes in developing a theory, Bohr found a trivial but noticeable error in one of the thought experiments Heisenberg had used to "prove" discontinuity (or rather, to disprove Schrödinger's continuity). Bohr wanted Heisenberg to recall the paper and fix it, but Heisenberg refused to do so. He stood his ground, even though he admitted later that he was reduced to tears by the pressure Bohr put him under to pull the paper before publication. According to historian of science Mara Beller, Heisenberg had his reasons:

> Heisenberg's adamant refusal to correct the mistake, despite Bohr's powerful opposition, is incomprehensible, unless we realize that Heisenberg had some vested interest in preferring a misleading description to the correct one. Heisenberg could not substitute Bohr's description for his own because the whole fabric of Heisenberg's argument against Schrödinger, based exclusively on discontinuity considerations, would collapse . . .[6]

The spring was particularly frustrating for Heisenberg because he felt he'd already said everything that needed saying in his uncertainty paper. As far as Heisenberg was concerned, they already had a complete theory—his. "We have a consistent mathematical scheme and this consistent mathematical scheme tells us everything that can be observed. Nothing is in nature that cannot be described by this mathematical scheme."[7]

The battle between the unrelenting Bohr and the tenacious Heisenberg raged through the halls of the institute. Oskar Klein, then 32, tried to mediate between the 41-year-old Bohr and the 25-year-old Heisenberg, without success. Bohr then sent for Pauli, but he couldn't come to Copenhagen at the time. Pauli had had a busy spring himself, creating matrices to incorporate non-relativistic spin into matrix mechanics.

In the middle of May, Heisenberg finally agreed to append a small note about the correction to the end of the March paper, but he refused to alter the paper itself.[8] He considered that he'd put out an excellent paper in the first place, and it was the time of the year when faculty committees were considering new appointments.

It must have been an enormous relief to Heisenberg when he found out early in June that he was once again on the short list for a position at Leipzig. His time at Copenhagen was to end in July anyway, and Oskar Klein would be taking over the job of Bohr's right-hand man. But Heisenberg didn't get just one job offer, he got three. Now he could afford to be choosy.

Pauli finally arrived in Copenhagen in mid-June to be the peacemaker—but not between Heisenberg and Bohr. Paul Hendry, historian of science, has suggested that an unspecified personal animosity had developed between Heisenberg and Klein in May, and it was this problem that Pauli successfully resolved.[9] Bohr and Heisenberg were still very much at odds with each other.

Heisenberg's departure from Copenhagen may well have been a relief to everyone who had to work closely with Bohr. Bohr's secretary noted disapprovingly that Heisenberg had not been very helpful to Bohr, nor was he as loyal as Kramers or as agreeable as Klein, and Heisenberg's tendency to "forget" to acknowledge the work of others had not gone unnoticed either.[10]

Schrödinger knew before he left on his American trip that he and Born were being considered for Planck's job in Berlin. An official decision had still not been made when he and Anny returned to Zurich in April. Schrödinger and Born were usually on good terms, even if they weren't close friends. They respected each other, despite Born's rejection of wave mechanics in favour of Bohr's version of the atom. But in the spring of 1927, Schrödinger was decidedly upset about the attacks on his theory coming out of Göttingen.

Jordan, in particular, had launched a full-scale assault on Schrödinger's credibility. In the spring, Jordan published a review of Schrödinger's papers in which he was quite scathing. Jordan asserted that it was unwise to expose physics students to wave mechanics until they had first received appropriate instruction in "physical matters in the Göttingen–Copenhagen spirit"; that the majority of physicists agreed that wave mechanics had no physical meaning; and that Schrödinger should have been satisfied to make a useful mathematical contribution to the Göttingen–Copenhagen physics and let it go at that. Schrödinger was outraged and complained to Born, who was still Jordan's supervisor, particularly since Jordan's claims were over the top and some of them were simply wrong. Born, however, demurred, excusing Jordan's excesses as youthful temperament.[11] Born was still carrying a lot of guilt over forgetting to submit Jordan's statistics paper, so he may not have been inclined to come down hard on his junior's unprofessional behaviour.

In early summer, the hiring committee for Planck's chair of theoretical physics made its decision. Born did not get the official "call"; the offer went to Schrödinger, and he accepted. The Berlin chair was one of the most prestigious positions in European physics, and it cemented Schrödinger's status in the physics community. In the end, it hadn't mattered that his best scientific work had been accomplished long after his thirtieth birthday.

Of course, the move to Berlin meant leaving behind his friends in Zurich, the mountains he loved and his darling Ithi, while Anny would be separated from Hermann Weyl. They were not likely to have difficulty making new friends, however, since Berlin society was just as permissive as Zurich society, if not more so. The appointment also meant he and Anny had to get busy packing so that he could be settled in Berlin in time for the fall semester, so Schrödinger didn't have much time available to continue working on the problems with wave mechanics.

The official list of speakers for the Solvay conference had been finalized, and Lorentz informed Schrödinger in June that he was on it. Lorentz wanted all the speakers to submit a written report of their presentation to him by September 1 so that copies could be made and distributed to the invited delegates in advance of the conference. With the move to Berlin, Schrödinger didn't expect he could have his report done until at least the middle of September.[12]

The Solvay scientific committee had agreed to keep the names on the list of speakers under wraps until after the conference to avoid any public demonstrations. Not all Belgians were keen to allow Germans into their country, and the ban against inviting German scientists to international conferences was only gradually being lifted. But the list sent to the speakers did, of course, name the others who would be making reports. That list included experimentalists Bragg and Compton, de Broglie, Heisenberg or Born, and Schrödinger. Kramers had been dropped. So had Einstein. In a letter to Lorentz on June 17, Einstein explained that he felt he had nothing of value to contribute that would be relevant to the current developments.

> [The] reason is that I have not been able to participate as intensively in the modern development of quantum theory as would be necessary for this purpose. This is in part because I have on the whole too little receptive talent for fully following the stormy developments, in part also because I do not approve of the purely statistical way of thinking on which the new theories are founded . . .[13]

Einstein *had*, in fact, been following the stormy developments. Heisenberg had been writing to him throughout the spring, arguing for the necessity of indeterminism in quantum theory. He had also challenged Einstein's desire for a deeper and more complete understanding of nature on the basis of the beauty of mathematical simplicity. "I do not really find it beautiful, however," he wrote to Einstein in June, "to demand more than a physical description of the connection between experiments."[14]

Bohr had also written to Einstein in April about the problems he was having trying to use the language of the classical world to describe the quantum world, and about his difficulties with the issue of visualizability. "If we only speak of particles and quantum jumps," said Bohr, "then a simple introduction to the theory, which is based on a consideration of the limitation of the possibility of observation, is difficult to find . . ."[15]

Einstein and Ehrenfest were also exchanging a flurry of letters in the spring, mostly about the Bose–Einstein statistics for an ideal gas, but some dealt with the implications of five dimensions.

Needless to say, Einstein was doing much more than letter writing. He too was pondering the puzzle of the wave-particle nature of the quantum world. Einstein had given a lecture in Berlin in February in which he'd stated that a true description of nature required neither a particle theory nor a wave theory. Rather, he said, "nature demands from us a synthesis of these two views which have thus far exceeded the mental powers of physicists." Einstein had said much the same thing in 1909, when he hoped there would eventually be a theory of light that would be a "fusion" of waves and particles.[16]

More important, in early 1927, Einstein began working on his first paper on quantum mechanics. Einstein's interest had been piqued by Bose's assertion in 1925 that light quanta in a gas could not be distinguished from each other. Here again was the puzzle of an unexplained influence on quantum particles, so that they did not seem to have true independence from each other as they would in the classical world. There was, Einstein said, "a mutual influence of the [quanta]—for the time being of a quite mysterious kind . . ."[17] This influence had appeared again in Schrödinger's wave equation in 1926, and forced Schrödinger to relocate his wave function ψ from classical space-time to configuration space (where degrees of freedom of movement replace spatial dimensions).

Einstein was attempting to provide a hidden-variables version of Schrödinger's work, much as de Broglie was doing, but he couldn't get around the "mysterious" way that particles seemed to be entangled.[18] Einstein submitted his paper for publication, and presented his work to the Prussian Academy of Sciences at the beginning of May. But, apparently uneasy about his inability to resolve the entanglement issue, Einstein withdrew the paper before it was published.

Without the pressure of writing a detailed report on quantum physics, Einstein was able to relax a little, but others could not. Heisenberg and Born started work on their joint report in Göttingen in early July, not long after Heisenberg had packed up and left Copenhagen. Lorentz had left it up to them to decide which of them would report matrix mechanics, but they'd opted to share the workload. Lorentz had also requested that they include Dirac's work in their report, and conveniently, Dirac was still in Göttingen when Heisenberg arrived, so the three of them could sort out what would go into the report. Born and Heisenberg divided up the workload so that each had about half the report to write, and agreed to get together again in August to finalize it.[19]

Bohr was working on a report as well, but not for the Solvay conference. He was scheduled to speak at a conference on physics in Como, Italy, in the middle of September. Bohr was pushing hard to get his talk completed. He wanted to be able to present a complete theory of the quantum world that would include stationary states, wave-particle duality and Heisenberg's uncertainty. But how to do it?

Oskar Klein accompanied Bohr and his family to their seaside holiday home in Tisvilde. A small cottage near the house served as an office for Bohr and his assistant, as well as any visiting scientists. During the whole summer Bohr talked through ideas and Klein took down the dictation. By July, Bohr had hit upon the idea of complementarity to describe the wave-particle nature of the quantum world. Both were necessary for a complete description of experimental results, yet they remained mutually exclusive—hence they were complementary to each other.

Bohr had lots of difficulty trying to formulate his ideas, especially when it came to incorporating Heisenberg's visualizable, particle-based uncertainty paper into a theory that also included Schrödinger's waves and Bohr's own belief in the non-visualizability of the workings of the atom. His draft notes provide evidence of his vacillations on the question of visualizability. Included in Bohr's drafts: "A few words about the relation with Heisenberg's work. Say that in essence the same endeavor. At the same time another spirit in that one stresses the visualizable and elementary character of the wave description."[20]

There is no doubt Bohr was suffering greatly with the mental challenge of trying to reconcile the seemingly irreconcilable and still develop a theory that made some kind of sense. According to Klein, Bohr's "suffering" that summer affected them all. "Bohr dictated and the next day all he had dictated was discarded and we began anew. And so it went all summer and after a time Mrs Bohr became unhappy . . . one time when I sat alone in the little room where we worked she came in crying . . ."[21] It is, perhaps, a measure of the tremendous difficulty Bohr was having that life in Tisvilde should deteriorate to such a level as to leave his adored Margrethe in tears.

It probably did not help when Pauli wrote to Bohr in August about de Broglie's recent paper, in which de Broglie had derived all the correct "pilot wave" equations, not only for a single particle but for a multiple particle system as well. Pauli suggested to Bohr that he might want to refer to the paper in his Como lecture. "De Broglie attempts here to reconcile the full determinism of physical processes with the dualism of waves and corpuscles . . . even if this paper by de Broglie is off the mark (and I hope that actually) still it is very rich in ideas and very sharp, and on a much higher level than the childish papers by Schrödinger . . ."[22] Given that Bohr was labouring mightily to integrate Schrödinger's "childish" work into a theory based on quantized stationary states, he was unlikely to be predisposed in favour of a classical theory on wave-particle duality that would so thoroughly undermine his own.

Bohr finally pulled his thoughts together sufficiently to write a short article on his new theory of complementarity, and it was sent off to the journal *Nature* on September 7.[23] The Como conference, where he would present his idea of complementarity for the first time, started five days later.

The talk in Como didn't go too well. In part, the scientists hearing this new idea had trouble following Bohr's reasoning, and Bohr was not a particularly good public speaker at the best of times. Historian of science Mara Beller has called Bohr's Como lecture "one of the most incomprehensible texts in twentieth-century physics."[24]

Heisenberg, who was at the conference, along with Born, Pauli, Schrödinger and Kramers—and about seventy other scientists— may have been surprised to hear that Bohr considered matrix mechanics to be a culmination of *Bohr's* research program. He credited his correspondence principle with being the bedrock upon which the successful new quantum mechanical methods were founded. Bohr played up the importance of Schrödinger's waves, and then poured cold water on the observables-only idea that had been a foundational principle of matrix mechanics, calling it inadequate since there was no point in applying space-time concepts to the quantum world anyway.[25]

It is unlikely that Heisenberg missed the barbs directed at him by Bohr—or the inclusion of concepts Bohr knew Heisenberg opposed—even if everyone else in the audience did. As an added touch, Bohr reinterpreted Heisenberg's idea of uncertainty to support his own notion of complementarity: one could not measure position and momentum simultaneously any more than one could measure waves and particles simultaneously.

Surprisingly, in the discussion following Bohr's talk, Heisenberg's comments were nothing but complimentary, but then he had three excellent job offers on the table. Later he admitted that he didn't believe complementarity was necessary, but since it didn't do any harm to his own interpretation, what did he care?[26] Born was also supportive, even with Bohr's backhanded swipe at Göttingen's version of matrix mechanics, but he had no fight left in him anyway; he was too tired.*

Bohr's confused theory obviously needed more work, so he and Pauli stayed behind in Como after the conference and spent a week reworking Bohr's report. After Bohr returned to Copenhagen, he informed the *Nature* editor that he would not be returning the proofs of his earlier article, but instead would be substituting the new and improved version when it was completed.

Bohr and Pauli had their own conflicts over how best to explain the quantum world. It was the visualizability thing again. Bohr had already abandoned classical four-dimensional space-time in order to integrate wave-particle duality, and visualizability could, in his opinion, occur only in classical space-time. Pauli disagreed. But they did agree that it was a good idea to elaborate on the integrated relationship between the observed and the observer—the thing being experimented on and the experimental apparatus—since the set-up of the experiment determined whether the wave or particle results appeared. With complementarity, there was no longer a classical independence between the observed and the observer.

When Pauli returned to Hamburg, he had more than just Bohr's theory of complementarity weighing on his mind. Pauli had been on the short list for a position at ETH in Zurich, but the official offer had gone to Heisenberg. However, Heisenberg had made no secret of the fact that he wanted an appointment in Germany, and was using his other job offers as leverage to negotiate the best possible deal for himself in Leipzig. That meant Pauli would still be in line for the Zurich position when Heisenberg officially turned it down.[27]

---

*For Born, the highlight of the Como conference was running into Ernest Rutherford and another scientist as all three were sneaking out of an unutterably boring presentation. They hired a taxi, toured around Lake Como and then had lunch together, the start of a useful professional friendship between Born and Rutherford.

Aside from career worries, Pauli was also deeply concerned about his mother. His father had taken up with a young sculptress and wanted to marry her, and Bertha Pauli was not taking the prospect of divorce very well. But it undoubtedly brightened Pauli's return to Hamburg to find an official invitation to the Fifth Solvay Conference in the mail. Lorentz had decided at the last minute that both Pauli and Dirac should be included among the conference delegates.[28]

The early fall of 1927 was a difficult time for Ehrenfest as well. Tatyana had gone to teach at a teacher's college in Russia. She hadn't left him, but she had chosen to utilize her considerable mathematical skills in the country of her birth. She had, after all, done her teacher training in St. Petersburg before going to university in Göttingen. But Paul was not used to being on his own, and he'd long relied on Tatyana to coax him out of his dark moods of self-doubt. Perhaps Tatyana's income would help ease the continual financial pressure Paul was feeling, but without her presence he had the added responsibility of handling both parental roles in the family.

Between his university responsibilities and his money-making efforts on the side, Ehrenfest was juggling a demanding workload. There was, after all, only so much money he could borrow from the well-to-do physicist and musician Adriaan Fokker, who had once been his assistant.[29] And Vassily's care in Jena was not getting any cheaper. It was little wonder that he despaired of being able to keep up with the rapid developments in quantum physics, and that he felt he was falling further and further behind. Tatyana's absence and the strain of trying to pay the bills and keep up with the new science were beginning to affect his professional life.

In the spring of 1927, Walter Elsasser and Robert Oppenheimer had both finished their doctoral degrees at Göttingen within two weeks of each other. Upon graduating, Oppenheimer headed off to the United States, where he had six offers of professorships to choose from. Elsasser, by contrast, had no job offers at all, and was facing the prospect of settling for a position as a high school science teacher. Then, out of the blue, Elsasser got a letter from Ehrenfest asking him if he'd be interested in being his new assistant. Samuel Goudsmit and George Uhlenbeck had graduated and were also taking positions at the same American university. Elsasser had no idea why Ehrenfest had chosen him since they'd never met, although he had seen Ehrenfest at scientific meetings. He assumed the recommendation must have come from Oppenheimer, who had spent time in Leiden with Ehrenfest the previous year. Mostly, however, Elsasser was just happy to have the possibility of an academic position, even if it was just for one semester.[30]

In August, Elsasser received a handwritten, six-page letter from Ehrenfest that rattled the 23-year-old scientist, who might have expected a discussion about physics at Leiden. "It dealt, instead," Elsasser wrote in his memoirs, "with some of Ehrenfest's psychological problems whose exact nature was unintelligible to me. I read the letter again, and then a third time, but was still unable to extract any sense from it."[31]

Uncertain what Ehrenfest expected from him, Elsasser nonetheless took the train to Leiden in September. Ehrenfest met him and once again poured out his personal difficulties. Elsasser had no idea how to handle this situation. He knew that Mrs. Ehrenfest was away teaching in Russia and that the two saw each other only occasionally, but he had little idea how he was supposed to respond to that. For whatever reason, Ehrenfest quickly took a dislike to Elsasser. "Within the first few days," said Elsasser, "I felt simply paralyzed from the shock his reception had administered. Within a very few weeks, his attitude changed from aggressiveness over indifference to outright hostility."[32]

This was not the Ehrenfest who went out of his way to encourage and help the young scientists he worked with. Elsasser was acutely aware of his shortcomings as a somewhat mediocre student, and that he was not living up to Ehrenfest's expectations. In early October, according to Elsasser, Ehrenfest walked into Elsasser's office and accused him of wearing perfume. "I will not tolerate perfume here," Ehrenfest shouted at him. "Get out. Go home, get out. Get out. Get out."[33]

Elsasser was dumbfounded, perhaps unaware of Ehrenfest's aversion to perfumes and strong scents, and he returned to his desk the next day as if nothing had happened. Ehrenfest had meant it, however, and informed Elsasser that his pay for the term would be forwarded and that he was to leave immediately. Elsasser was greatly dismayed. Being fired from his first position as an assistant was a death knell for his hopes in academia, and with the increasingly hostile environment towards Jews in Germany his job prospects were almost non-existent. Now he was facing an emotional breakdown of his own.[34]

Whatever the inner turmoil that had seized Ehrenfest in early October, he'd snapped out of it by the time he arrived in Brussels at the end of the month for the Solvay conference.

Chapter Sixteen

COMING UNDONE

First they told us the world was flat. Then they told us it was round.
Now they are telling us it isn't even there.

IRVING OYLE,
physician and author

The plans for the Fifth Solvay Conference did not come together without a few snags. One was the overlap between the dates for the conference and the centenary in Paris for the mathematician and physicist Augustin Fresnel. In the 1820s, Fresnel's experimental work convinced him of the wave nature of light, as opposed to Newton's belief that light was a stream of corpuscles. A century later the light-is-a-wave or light-is-a-particle debate was still going strong.

The Fresnel celebrations were scheduled to open on Thursday, October 27, right in the middle of the Solvay conference. It was simply impossible for Hendrik Lorentz to reschedule the conference. The rooms had been booked at the Hôtel Métropole for all the invited scientists, the catering had already been arranged for their meals to be served at the Institut de Physiologie, and a grand reception, hosted by Armand Solvay, was arranged for the last day of the conference. It was too late to reschedule anything.[1]

In another scheduling conflict, de Broglie had agreed to give a lecture to the Société de Physique on the twenty-seventh, in part because he had a somewhat distant family connection with Fresnel,* as well as a professional interest in light waves. Lorentz and the Solvay organizing committee decided it was best to leave the schedule open for Thursday and extend a general invitation to the Solvay guests to also attend the Fresnel celebration. Since the Fresnel event didn't begin until the evening, the guests could travel to Paris during the day and return the next morning. The Solvay committee was covering all their expenses in any event.

The Fifth Solvay Conference got under way at ten o'clock on Monday morning, October 24, 1927, with Lorentz as the chair. The first morning and afternoon lectures were given by Lawrence Bragg and Arthur Compton, the only talks that remained from the original agenda. The Free University of Brussels hosted a reception Tuesday morning, and de Broglie gave his presentation on his "double solution" in the afternoon. Werner Heisenberg and Max Born were up Wednesday morning, with Erwin Schrödinger in the afternoon. Thursday was left open so that those who wanted to could travel to Paris. Most did, except for Ehrenfest, Dirac, Einstein, Schrödinger and a few others. Friday and Saturday were reserved for general discussions but included a Saturday lunch hosted by the Belgian king and queen and the banquet hosted by Solvay to wrap up the gathering on Saturday night.[2]

It wasn't a secret in the physics community that Bohr and Einstein had their differences—the light quanta dispute raised by the BKS paper had made that clear—but anyone expecting fireworks between the two during the conference would have been quite disappointed. They did get into a debate over a number of points in the Bohr theory, but it was largely a private discussion witnessed primarily by Ehrenfest, not unlike the debate that Bohr and Einstein had had two years earlier in the privacy of Ehrenfest's home. Heisenberg did overhear some of the discussion between Bohr and Einstein as they walked together from the hotel to the conference

---

*Augustin Fresnel was the son of the second duc de Broglie's goddaughter.

building. But the philosophical discussions between Bohr and Einstein didn't seem to hold much interest for the others who might be sitting with them at meals.

For Heisenberg, quantum mechanics had been completed when he produced the uncertainty paper in the spring. "I thought that now we have a mathematics scheme which is consistent, it can either be right or wrong, but if it's right then anything added to it must be wrong because it is closed in itself."[3] The belief that quantum physics was complete was buttressed by a similar argument from Dirac when he had produced his transformation theory in 1926 as the mathematical bridge between particles and waves. According to Dirac, everything that needed to be said about the quantum world could now be said, albeit in abstract mathematical formalism.

But Einstein and Bohr had lots more to say. They wanted to *understand* the quantum world. Bohr was still struggling with the rewriting of his paper on complementarity that he'd promised the editor at *Nature*, a revision of the talk he'd given at Como. And Einstein was battling the same problem he'd got tangled up with in the spring with his first paper on quantum mechanics. In the quantum world, it didn't seem possible to separate particles so that what happened to them had a clear cause and effect. The point of introducing "ghost fields," "virtual fields" and "pilot waves" was to give some kind of causal explanation for the motion of particles where there didn't seem to be one.

Bohr had been puzzling over the same thing, and his solution was complementarity. The BKS paper had failed because he had assigned distinct and separate states to particles, which meant that energy and momentum could not be conserved. But since experimental evidence showed they were indeed conserved, it must mean that particles really did have some sort of dependence on each other. This was what Einstein had called the "mysterious" mutual influence of particles in the quantum world.[4]

Bohr was happy with complementarity as an explanation for the mutual dependence problem. It covered other paradoxes as well.

Position and momentum were both required for a physical description, but, according to Heisenberg's uncertainty, both could not be measured accurately at the same time. The interior of an atom could be described in classical space-time or it could be deterministic, but it couldn't be both at the same time. Similarly, experiments could be set up to test for waves or particles, but not both at the same time. Complementarity didn't explain *why* certain properties in the quantum world seemed to be inextricably linked, just that they were. By declaring the quantum world non-visualizable, Bohr had done away with the need to explain the paradoxes.

For Einstein, however, Bohr's complementarity closed the door on any possibility of understanding the mutual dependence thing, whatever it was. It rendered the quantum world unknowable, and forever incompatible with the classical world of general relativity. For his part, Einstein was not ready to give up hope of ever understanding this strange property thrown up by the microscopic world.

Each tried to persuade the other to his point of view, and when Bohr needed someone to help him work through a thought experiment or question posed by Einstein, he'd show up at Ehrenfest's door. Sometimes Bohr would just pace and think, saying almost nothing. "Every night at 1 a.m.," Ehrenfest wrote, with lots of exclamation marks and words "shouted" in full capitals in a letter to George Uhlenbeck and Samuel Goudsmit, "Bohr came into my room just to say ONE SINGLE WORD to me, until 3 a.m."[5]

Ehrenfest thoroughly enjoyed being at the centre of such a debate, and he conveyed his intense excitement in his lengthy letter to his two former students. Ehrenfest sometimes played the role of mediator between his two friends, lightening the mood with a joke when things got too intense. Given that Ehrenfest was enormously frustrated by his inability to get a handle on quantum physics, what better way for him to learn than to engage in just the kind of debate he particularly relished with Bohr and Einstein.

None of what Bohr and Einstein were debating in private was said during the daytime formal presentations and discussions. The two men contributed almost nothing to the lengthy debates that followed each presentation. There was certainly plenty to debate, since three distinct methods of describing quantum phenomena had been floated: de Broglie's classical pilot wave theory, which sought to integrate wave-particle duality via pilot waves; Schrödinger's semi-classical wave mechanics; and Heisenberg and Born's non-classical, probabilistic quantum mechanics (a blend of wave and matrix mechanics).

The discussion following de Broglie's presentation was quite lengthy, with questions from Lorentz, Born, Kramers, Pauli and Ehrenfest focusing mostly on his novel use of trajectories, but it was Schrödinger who challenged de Broglie's version the most, perhaps because he wanted to distance himself from the man who had created the ideas upon which his own theory was based. According to the official records, the only comments following the Heisenberg and Born presentation were from Dirac, promoting the use of matrices. Neither de Broglie nor Schrödinger challenged the Heisenberg–Born presentation, not even after they announced that their version meant quantum physics was finished and complete.[6] "We regard quantum mechanics as a complete theory for which the fundamental physical and mathematical hypotheses are no longer susceptible to modification . . ." declared Heisenberg and Born. "Our fundamental hypothesis of essential indeterminism is in accord with experiment. The subsequent development of the theory of radiation will change nothing in this state of affairs."[7]

Because the official record of the conference was translated from German into French and essentially sanitized, removing any emotion or personal invective, the discussion periods come off sounding formal and stilted, with nary a stumble or misspoken word. That's entirely natural, since each speaker was given the opportunity to edit or revise his comments prior to publication of the proceedings. But even in the tidied-up version, a hint of emotion leaks through in the discussions that followed Schrödinger's presentation. A number of physicists questioned Schrödinger, but

Heisenberg and Born really drilled into the weak spot in wave mechanics, which Schrödinger was well aware he had not yet resolved: the inability of wave mechanics to produce correct solutions in three spatial dimensions.

It wasn't until near the end of the conference, during the general discussions, that Bohr was given an opportunity to explain his complementarity theory. Neither Heisenberg nor Born actually agreed with Bohr's theory—not that they were letting on at the conference, though. Bohr, Heisenberg, Pauli and Born backed each other up during the discussions, as did the ever-loyal Kramers. Born provided the same unquestioning support for Bohr that he'd exhibited at Como when, after Bohr's confusing talk, he announced, "Herr Professor Bohr has presented the views that we have formed about the basic concepts of quantum theory in such an appropriate manner that there is nothing left for me to do but add a few remarks."[8] Born had little fight left in him; it was just easier to go along with Bohr. As for Heisenberg, he could support Bohr publicly because it didn't cost him anything. He had the job he wanted, and besides, he was just a bit ashamed of some of the things he'd said to Bohr earlier in the year in the heat of their arguments.

There may not have been much in the way of heated debate during the discussions that followed the reports or in the general discussion later, but tension there certainly was. Physicist Irving Langmuir, a former Göttingen student who had become a researcher for General Electric in the United States, had brought with him a hand-held movie camera and filmed two quite different scenes at the conference.[9] The first is a laughing, chatting group outside the conference building—Heisenberg grinning for the camera, Ehrenfest laughing and then sticking out his tongue, Born mugging a bit as well, Dirac and de Broglie looking shy but relaxed, and Bohr and Schrödinger talking animatedly but with easy smiles between them. (Pauli, on the other hand, looks dour and quite unimpressed with being filmed.) The second scene, in marked contrast to the first, shows the participants exiting the physiology building looking grim and tense. As they stream out of the doorway—Lorentz, Planck, Einstein, Bohr, Schrödinger, Born and the rest—hardly any of them has a smile to spare for Langmuir's film.

The tension that was captured in the short film finally exploded when Einstein stood to speak on the last day of the conference, after Bohr had explained his theory of complementarity to the ensemble of scientists. Einstein gave complementarity the thumbs-down, which certainly would have come as no surprise to Bohr. But it got the others all excited. Bohr biographer Ruth Moore described the scene in the salon:

> A dozen physicists were shouting in a dozen languages for the floor. Individual arguments were breaking out in all parts of the room. Lorentz, who was presiding, pounded to restore order. He fought to keep the discussion within the bounds of amity and order. But so great was the noise and the commotion that Ehrenfest slipped up to the blackboard, erased some of the figures that filled it and wrote: "The Lord did there confound the language of all the earth." As the embattled physicists suddenly recognized the reference to the confusion of languages that beset the building of the tower of Babel, a roar of laughter went up. [10]

Einstein also had a problem with how the Copenhagen–Göttingen school used probabilities to determine how likely it was that a wave would throw up a particle at a certain point. Since the probability wave was continuously distributed through space, Einstein argued, why couldn't it throw up several particles in different places at the same time? One would have to assume, he said, that there was "an entirely peculiar mechanism of action at a distance" to prevent the rest of the wave from simultaneously producing another particle somewhere else, and therefore the wave function could not be a complete description of what was going on. "In my opinion," concluded Einstein, "one can remove this objection only in the following way, that one does not describe the process solely by the Schrödinger wave, but at the same time one localises the particle during propagation. I think that Mr. De Broglie is right to search in this direction."[11] Otherwise, he added, the probability interpretation contradicted the tenets of relativity.

Bohr responded by saying that he didn't know what Einstein was talking about. "I feel myself in a very difficult position because I don't understand what precisely is the point that Einstein wants to [make]." Bohr nonetheless agreed that the Schrödinger wave did not give a complete description—but then his own theory was based on using both waves *and* particles to describe the quantum world, as much in opposition to Heisenberg's particles-only theory as to Schrödinger's waves-only version. "Using a rigorous wave theory, we are claiming something which the theory cannot possibly give."[12] But that didn't mean Bohr was going to support de Broglie's theory, even if it also provided for both waves and particles.

The conference delegates were not having an easy time of it, and not just because of the science. Language was an issue as well, and in more ways than one. Only a few of the conference guests were multilingual. English experimentalist W.H. Bragg, Lawrence's father, declined an invitation to attend, even though he was on the organizing committee, because he found it so difficult to follow the discussions in French and German, despite Lorentz's best efforts to translate.[13] Lorentz also had the challenge of translating Bohr's convoluted statements, which were not particularly easy to comprehend even if one was fluent in the language he was using. In the letter to Uhlenbeck and Goudsmit, Ehrenfest described "poor Lorentz as interpreter between the British and the French who absolutely did not understand each other. Summarising Bohr. And Bohr responding with polite despair."[14] It's no wonder that Paul Langevin described the Fifth Solvay Conference as a gathering where "the confusion of ideas reached its peak."[15] There were, after all, three distinct versions of quantum physics on offer at the conference, each one able to explain most—but not all—quantum phenomena. They were theories in flux, even if, during their joint presentation, Born and Heisenberg made a bold declaration of completion.

Towards the end of the conference discussions, Lorentz offered a gentle reminder to those who were taking up dogmatic stands based on theoretical preferences. "I happily concede that the electron may dissolve into a cloud. But then I would try to discover on which occasion this transformation occurs. If one wished to forbid me such an enquiry by invoking a principle, that would trouble me very much." He also cautioned against being too quick to dismiss the value of visualizability. "I should like to preserve this ideal of the past, to describe everything that happens in the world with distinct images. I am ready to accept other theories, on condition that one is able to re-express them in terms of clear and distinct images."[16]

There was no clear winner or loser in the discussions, although the fact that the Copenhagen and Göttingen scientists kept endorsing each other might have made it seem as if they had the better argument. Whatever their animosities, hurt feelings or grievances with each other—or even their strong disagreements—they stood united behind Bohr. They were conscious of carrying the weight and credibility of two of the world's leading schools in quantum physics— the Bohr temple and the Born mathematical physics dynasty.

Still, they did not "win" the day; there was no vote taken at the end of the conference to see which theory was most popular. Years later Heisenberg suggested that the most important success of the Solvay conference was the way "we" were able to stand against any challenge to the Copenhagen view. And who were "we"? As Heisenberg recalled, "I could say at that time it was practically Bohr, Pauli and myself. Perhaps just the three of us."[17] Heisenberg could make such a claim only by glossing over the fact that Bohr, Pauli and he had yet to reconcile their own differences on such issues as visualizability, never mind reconciling their various differences with Dirac, Born and Jordan.

Heisenberg, it seemed, felt quite good about how it had all gone. He sent off a victorious letter to his mother and father. "I am satisfied in every respect with the scientific results. Bohr's and my views have been generally accepted; at least serious objections are no longer being made, not even by Einstein and Schrödinger."[18]

Einstein and Schrödinger would have been astonished to hear that they had no "serious objections."

De Broglie and Schrödinger may have felt more than a little chastened by the concerted efforts of the Copenhagen and Göttingen schools to challenge their theories. Granted, both theories still had problems—but then so did Bohr's.

As everyone headed home after the conference, the landscape of physics that had generated the three different theories of quantum physics was already shifting. The "quantum ten" quickly scattered as former alliances and working relationships changed. Some, such as Heisenberg, Pauli and Dirac, were moving up the professional ladder and had new responsibilities to handle; they were no longer the bright young men of the "boys' club" trying to make a name for themselves in physics.

Einstein and de Broglie travelled on the same train after leaving the conference. Einstein was weary, but he took time to reassure de Broglie about his pilot wave theory. "Carry on," he said. "You are on the right road."[19] De Broglie, however, did not take Einstein's advice. Perhaps he didn't feel that his solitary effort could stand in the face of the Copenhagen juggernaut, and upon his return to Paris he gave up working on his pilot wave theory.

Schrödinger was not so easily cowed. He'd already gone head-to-head with Bohr, and withstood the sniping and low blows from Heisenberg and Jordan. But he had a new job in Berlin, and the winter lecture semester started two days after the Solvay conference ended.[20] He missed Ithi and was already pondering the possibility of arranging trysts. They were not lovers yet, but Schrödinger was hoping it would happen soon. Of course, that wouldn't stop him from enjoying the pleasures of the flesh in Berlin betwixt liaisons with Ithi.

Heisenberg too had to get to work after the conference. His appointment as the chair of theoretical physics in Leipzig had been finalized just before the conference. He would now have his own pulpit from which to preach the Copenhagen gospel—or at least his version of it. And as part of the negotiated deal, Heisenberg would be allowed an eight-month leave of absence to take the new gospel around the world. But first he had to get settled in Leipzig.

Heisenberg's new home was a comfortable apartment on the top floor of the institute in Leipzig, an old building that was, ominously, bordered by the garden of a mental institution on one side and a graveyard on the other.[21] No doubt it took some heavy lifting by somebody—or several somebodies—but before long his apartment boasted a piano as well, and students could often hear the faint sounds of beautifully played sonatas drifting down the quiet hallways.

Since Heisenberg had turned down the ETH position in Zurich, Pauli had good reason to expect a call from them. But the call that came on November 12, just two weeks after the Solvay conference ended, was to inform him that his mother had taken an overdose of the barbiturate Veronal. She was rushed to hospital, but doctors were unable to revive her. She died two days later.[22] Pauli was devastated. He'd been close to his mother, and her death sent him into a tailspin. He had always enjoyed his drink, but now it took on a harder edge. He had his own dark moods to contend with.

In December, Pauli finally got the awaited call from ETH: he would become a full professor of theoretical physics in Zurich. Pauli thereafter signed himself as Wolfgang Pauli, not Wolfgang Pauli Jr., perhaps because of his "graduation" to professor, but perhaps also as a means of distancing himself from his father.[23] Pauli started at Zurich in February, but his drinking began getting in the way of his job. He stopped working on quantum physics again, his lectures were confusing and poorly prepared, and his trademark sarcasm had become more mean than witty. It only got worse when his father married his mistress with unseemly haste, only six months after his mother's suicide.

As soon as Dirac returned home from the conference, he squirrelled himself away in his rooms at Cambridge to work on one of the unsolved problems in the allegedly "completed" quantum mechanics. Nobody had been able to figure out how to incorporate the "spin" of an electron with the relativistic wave equation. Pauli and Schrödinger had been working on it during 1927, and were inclined to think that it would take some kind of sophisticated model of the electron that no one had yet come up with. Dirac disagreed—he figured the relativistic spin should fit naturally into quantum physics. But he was working on the problem secretly. Dirac was well aware that most of the work he'd done so far in quantum physics had also been done by others, such as Born, Jordan or Pauli, who published before he did. He felt that he hadn't yet contributed anything original that would bring him out from under the shadow of the Göttingen and Copenhagen schools, so this time he didn't tell the others what he was working on.[24]

Initially, Dirac assumed that Schrödinger had abandoned his first relativistic equation too quickly back in the winter of 1925 because it didn't produce solutions matching experimental results. He had been on the right track, thought Dirac, but had backed off out of fear that a failed equation would undermine his entire wave theory.[25] Dirac had always been interested in general relativity— it's what he went to Cambridge to study in the first place—so he revamped Schrödinger's original equation into a highly mathematical approximation that incorporated relativity and matrices. His different approach, he hoped, would solve the problem that the flood of relativistic wave equations produced the previous fall had not.

Unfortunately, his own approximation seemed a bit off, but, taking a lesson from Schrödinger's earlier timidity, Dirac quickly wrote up the paper in January with what he had. He wanted to win the race for publication this time, and he did.[26] His paper was published in January—beating out Kramers in Utrecht, Pauli in Hamburg and Jordan in Göttingen—and it enjoyed an enthusiastic reception from Bohr and Born. Dirac was elected a fellow at Cambridge; he had finally emerged from the shadow of the others. He would henceforth cast a shadow of his own—albeit a very long, thin one.

In Copenhagen in early 1928, Bohr and Oskar Klein were still working on the complementarity paper Bohr had promised the *Nature* editor the previous fall. In December he'd again written to the editor to apologize for the delay. "I have found it most expedient to rewrite the whole article . . . I am much ashamed for all the trouble I am giving you."[27]

After the Solvay conference, Lorentz still had to write up his own report from the Como conference in September, so he assigned the responsibility of preparing the draft notes of the Solvay proceedings to the scientific secretary. The notes were to be typed up and sent to the speakers to help them write their reports for a book on the conference.[28] In the spring, the scientific secretary in charge of publishing the proceedings was waiting on Bohr as well; the Dane still had not returned the notes on his contributions.

Lorentz never did complete his Como report. He fell ill, and on February 4, 1928, he died. The entire nation of Holland was immediately in mourning. Flags hung at half-mast at the railway station, the cavalry barracks and the town hall in Haarlem, and a special train brought students and colleagues to the city for the funeral,[29] just as a special train had transported guests on the same trip in 1925 for Lorentz's fiftieth-anniversary celebrations. At the stroke of noon on February 10, the day of the funeral, the Dutch telegraph and telephone services fell silent for three minutes in tribute to the great physicist.[30] Einstein, Ehrenfest, Langevin and Rutherford, representing their respective countries of Germany, Holland, France and Great Britain, each spoke at the graveside.

Einstein wasn't in particularly good shape himself. He was scheduled to give the inaugural lecture for the Alpine University in Davos, Switzerland, where a large group of prominent European academics were starting a month-long set of lectures. Davos, with its altitude of five thousand feet above sea level, was the site of a number of sanatoriums for the treatment of tuberculosis. The objective of the new university was to hold lectures and discussions that would, hopefully, brighten the lives of the young people from the sanatoriums. About 45 leaders in science, philosophy, literature, law and sociology were scheduled to give six or seven lectures a day to roughly 360 students and 400 other attendees.[31]

Einstein opened the March lecture sessions with a talk on "The Fundamental Concepts of Physics in Its Development," a timely topic hard on the heels of the Solvay conference. Einstein got right into the spirit of things. According to historian of science Gerald Holton, "He attended other professors' lectures assiduously, met with individuals and small groups, even played his violin at a chamber concert, to help raise funds for this new university."[32] In the process, Einstein wore himself out. He went to stay with a friend at his Swiss mountain chalet, and insisted on carrying his own bag as he and his friend trudged up the slope in the deep snow.[33] Einstein collapsed in the snow, his host frightened that he'd had a heart attack. He hadn't. He was later diagnosed with an enlarged heart and high blood pressure, which, no doubt, had been exacerbated by the thin mountain air.

Einstein was bedridden for several months. At one point he feared he might be at death's door, and started making arrangements in the event he did not survive. He did survive, but he had little energy to spare for the still-unresolved problems of quantum physics.

Neither did Max Born, whose workload at Göttingen had not diminished. His home life had become much quieter, though. His older daughter Irene had been sent off to boarding school in Switzerland, and in the spring Hedi had moved out of the house.

She said she needed to be alone for a while. By the summertime Hedi had moved back home, but it wasn't long before she went off again. This time she wasn't alone. She'd fallen in love with Gustav Herglotz, a mathematician at the university whom Max sometimes worked with, and one of Paul Ehrenfest's long-time friends from their university days in Vienna.[34]

Max's worst fears had been realized, but he soldiered on, juggling his many administrative duties and taking on too many doctoral students. By the fall he was in bad shape. At one point in October, a dismayed seven-year-old Gustav found his father sobbing in the bathroom. Max was fast approaching a nervous breakdown. By the end of December he was in a sanatorium in Switzerland, where he wrote to a friend, "It has not gone well for me. I have over-stressed my nerves."[35]

The year had not gone particularly well for de Broglie either. He was heavy-hearted about the failure of his pilot wave theory and considered himself a failure, as did his mother. When the de Broglie matriarch died in 1928, as Louis's sister Pauline later wrote in her memoirs, she died "taking to her grave an image of him as a failure, a ne'er-do-well who will never give her the posterity she was longing for."[36]

The de Broglie family sold the grand mansion in Paris, something Louis had been expecting would happen once his mother died, and he moved into a smaller house in a quiet neighbourhood, along with a couple of servants.[37]

Scientists and students arriving at Copenhagen in the winter of 1927–28 expecting to work with Bohr on the exciting new physics must have been quite disappointed. Bohr was working on complementarity, and that was philosophy, not physics.

It took Bohr until the end of March to finally complete his paper on complementarity, "under strong pressure" from his brother Harald to finish with it.[38] It was not really the "Como paper" anymore, since it had undergone extensive revisions and a great many rewritings since the previous summer. Bohr's obsessive rewriting had also been holding up the editing of the Solvay proceedings. He laboured mightily over everything he wanted to say, and it would have been out of character simply to rewrite the partial sentences and incomplete thoughts that are typical of recorded discussions. He even went to Utrecht in early March to work with Kramers on the notes. Finally, he requested that the scientific secretary simply substitute his carefully crafted, soon-to-be-published *Nature* paper in lieu of the lengthy discourse on complementarity he'd given during the general discussions.[39]

When the proceedings volume was finally published in December 1928, it contained Bohr's *Nature* article, printed alongside the official reports, leaving the impression that he'd been one of the invited speakers at the conference, which, of course, he was not.

The year following the Solvay conference was a turbulent one. The creators of quantum physics had dispersed. Bohr was still heading up his own "school," but he had not been doing actual physics for several years; he wasn't able to follow the necessary mathematics. He was slowly becoming less and less of a physicist and more and more of a philosopher. But the Bohr temple was still going strong, with its emphasis shifting from the Bohr atom to the new mysteries of the atomic nucleus.

For his part, Born had temporarily ceded the helm of the Göttingen mathematical physics dynasty after his breakdown, because it was going to take some time to mend his shattered nerves. In Munich, Sommerfeld's "nursery" was still producing budding wunderkinder, but Sommerfeld was greatly dismayed at the growing National Socialist presence on campus. The Nazis had found fertile ground for their recruiting among university students, who would wear swastika armbands and make life difficult for Jewish professors and students.

Schrödinger was still routinely distracted by his "second occupation," as he referred to his love life, anticipating the final seduction of the lovely Ithi after her seventeenth birthday in the summer. Meanwhile, Jordan had finished his habilitation and had taken the position that opened up in Hamburg when Pauli moved to Zurich, although his unremedied stutter made it difficult for him to lecture.

Born and Einstein had both been sidelined by health problems, and the problems with quantum physics, particularly the interpretations, remained unresolved. But that didn't stop Heisenberg from remembering the Solvay conference in a rather different light. "Through the possibility of exchange between the representatives of different lines of research," he said in 1929, "this conference has contributed extraordinarily to the clarification of the physical foundations of quantum theory; it forms so to speak the outward completion of quantum theory . . ."[40]

Chapter Seventeen

PICKING UP THE PIECES

*The fact that an opinion is widely held is no evidence
whatever that it is not utterly absurd; indeed in view
of the silliness of the majority of mankind, a widespread
belief is more likely to be foolish than sensible.*

BERTRAND RUSSELL

When Albert and Elsa Einstein sailed for the United States in December 1932, Albert told Elsa they might not be able to go back again. He was prescient, seeing past the disbelief of many of his contemporaries that a civilized country like Germany would long tolerate a leader like Hitler. The political climate in Germany was becoming decidedly unhealthy for people who opposed the agenda of the National Socialist German Workers Party, and particularly for outspoken Jews such as Einstein.

Anti-Jewish sentiment in Germany had calmed a little in the latter part of the 1920s, once the Weimar government finally got inflation under control. But the great stock market crash in 1929 had triggered a worldwide economic depression, and the rising unemployment and financial troubles in Germany once again inflamed nationalist feelings. In the national election in the summer of 1932, the National Socialist Party consolidated its political position by winning a large number of seats, although not enough to take power. The Weimar government was on shaky ground, with different factions jockeying for control.

Einstein could see the writing on the wall. He had informed Planck that he had accepted a position at the newly created Institute of Advanced Studies at Princeton University, intending to spend half the year in Berlin and the other half in Princeton. But first he had to leave Germany in December to fulfill his agreement to spend a third and final winter at Caltech in Pasadena.

Einstein's travel schedule didn't prevent him from pursuing a unified theory, but to do that he had to develop a theory of the quantum world that was about physical reality and not just observational results. In January 1929, the year of Einstein's fiftieth birthday, word got out that he was working on a new theory. According to biographer Philipp Frank, the idea of a new work by Einstein stirred up considerable excitement. He was, after all, an international celebrity.

> To the public at large it seemed to be an especially attractive idea that on the very day on which he attained fifty years, a man should also find the magic formula by which all the puzzles of nature would be solved . . . Hundreds of reporters besieged his house. When some reporters were finally able to get hold of him, Einstein said with astonishment: "I really don't need any publicity." But everyone expected some new sensation that would surpass the wonder produced by his previous theories.[1]

The newspapers that were so eager to write about Einstein's new theory would have to wait until it was printed in an academic paper, but that didn't stop the headlines. The *New York Times*, for instance, exclaimed breathlessly: "EINSTEIN ON VERGE OF GREAT DISCOVERY; RESENTS INTRUSION" and "EINSTEIN IS AMAZED AT STIR OVER THEORY. HOLDS 100 JOURNALISTS AT BAY FOR A WEEK."[2]

When the new theory finally came out, it was almost entirely complicated mathematical formulas. To the layperson, said Frank, it would have been on par with an Assyrian cuneiform inscription.[3] Nonetheless, the public interest in Einstein's new theory was so great that Selfridges department store in London pasted six pages of newspaper coverage in its window so that people passing by could read all about it.[4]

It wasn't Einstein's first attempt at a unified theory, and his scientific colleagues were not as impressed by his unfathomable "Assyrian cuneiform" as the general public. Einstein soon discovered problems with his latest unified theory and got to work on another one. Pauli dismissed Einstein's repeated efforts as providing "on average, one theory per annum . . ."[5]

Einstein hadn't had much time to visit Leiden, but he spent some time there with Paul Ehrenfest in April 1930. The Ehrenfest household was no longer the happy home Einstein had so much enjoyed. As Einstein left Europe at the end of 1932, he hoped his efforts to help heal the rift between Paul and Tatyana had been of some help. That summer, when they were still separated, he'd gone to see them both in Leiden, but it wasn't clear how much good it had done.[6]

With Tatyana teaching in Russia so much of the year, Paul had been floundering on his own. In the fall of 1931 he wrote a plaintive letter to his brother Hugo, a doctor in St. Louis, Missouri. Hugo was, of course, quite familiar with his little brother's shaky sense of self-worth, since his black moods had been going on since he was a teenager. Hugo wrote back, rather bluntly, "You are depressed, suffer from a physical inferiority complex, get on your best friends' nerves, whenever you are together for a long time . . . Completely incorrectly and egotistically you express the words, 'Nobody loves me.' If you were only a bit more considerate you would easily find that everybody loves you."[7]

Paul needed somebody to love him, and in 1932 he turned to another woman for the comfort and nurturing he craved. Nell Posthumus-Meyes was an art critic and a visibly warm person who was quite unlike the independent and mathematically minded Tatyana.[8] Paul had never imagined himself to be the kind of man who would cheat on his wife, and he felt terribly guilty about what he believed he'd done to his family. He had no idea how he could possibly afford to divorce his wife and continue to support her while supporting Nell as well. And then there was the cost of Vassily's care in Jena.

Ehrenfest, who had never really believed he had been a worthy replacement for Lorentz, had been looking for a job in either the United States or Russia. He'd made a second trip to America in 1930, hoping to generate some job offers, but nothing had materialized. In 1931, Ehrenfest wrote to Niels Bohr about his failure to keep up with the new developments in physics. "I have completely lost contact with theoretical physics. I cannot read anything any more and feel myself incompetent to have even the most modest grasp about what makes sense in the flood of articles and books. Perhaps I cannot at all be helped any more."9

With so many scientists, especially Jewish ones, looking to escape the ugly climate in Germany, Ehrenfest felt he should vacate his position in Leiden and open it up for someone more worthy. According to historian of science Martin J. Klein, "He knew he was widely recognized and honoured as a teacher. His students worshipped him and thought him very effective as a teacher and with their research, even if he didn't believe he was much help to them in their research."10

Paul and Tatyana talked of divorce, but she was reluctant to consider it. Paul was very distraught about the whole situation, and he had yet to see any light at the end of the tunnel. He'd not got any sign of interest from any American or Russian university. "He'd received no offers," says Klein, "and was in the depth of depression. The future seemed very restricted to him. At some point, he made the resolve to end his life and the more horrible decision to end the life of his son."11

In mid-August 1932, Ehrenfest sat down and wrote two suicide notes, one addressed to friends such as Bohr, Einstein, James Franck and Gustav Herglotz; the second addressed to some of his former students. In the letter to his friends, he said that his life had become unbearable, he simply had to vacate his position in Leiden, and he'd didn't seem to be able to follow the developments in physics no matter how hard he tried. "This made me completely 'weary of life,'" he wrote. "I did feel 'condemned to live on' mainly because of the economic cares of the children . . . I have no other 'practical' possibility than suicide, and that after having first killed [Vassily]."[12]

Ehrenfest didn't send the letters, however, and seemed to set aside the idea of suicide. Perhaps Einstein's intervention had been of some help after all. Talk of divorce was also shelved, but the situation remained unresolved when Tatyana left again for Russia.

In the fall, Ehrenfest received notification from the facility in Jena, where Vassily was being cared for, that the passport under which Vassily had been admitted had expired in 1924, and because of the new strictures on foreigners the boy needed new documentation.[13] With the political situation in Germany getting more and more uncertain, Ehrenfest decided it would be a good idea to bring Vassily back to Holland, although the move would likely be disruptive for the fourteen-year-old boy.[14] There was a large facility in Amsterdam, the Professor Watering Institute, and by January 1933, Vassily was safely settled there.

The Ehrenfests weren't the only ones talking about divorce; so were Max and Hedi Born. Hedi's affair with Gustav Herglotz had been the talk of the town in Göttingen in 1929. Paul Ehrenfest arrived in July that year for his usual summer visit, but instead of staying with the Borns, as he usually did, he stayed with Herglotz. Hedi, who was still living at home, would walk over and have dinner with them, and Ehrenfest, who was not yet having an affair of his own, firmly told Hedi to get things sorted out. After Ehrenfest returned to Leiden, Hedi told Max she wanted a divorce so that she could marry Herglotz. Max refused.[15]

It had taken Max several months after his breakdown to feel strong enough to return to the university, but even then he did not resume his lecturing and supervisory roles. Instead, he spent most of his time at home, writing a book on quantum physics in collaboration with Pascual Jordan. Students who arrived at Göttingen hoping to work with Born had to find someone else. With the added pressure of a possible divorce, Born took a two-month leave from the university and went away with the children on holiday. Hedi stayed home with Herglotz.[16]

When Max returned, Hedi left for a rest cure at a sanatorium. She had ended her affair with Herglotz. After she finally returned to Göttingen, she organized the attic rooms into an apartment so she could be by herself. Max kept his head down and went on working on his book. They didn't get divorced, but they were not happy.

Born's book, *Elementary Quantum Mechanics*, was published in 1930. It had taken longer than expected to write, with Jordan in Hamburg. The idea had been to write a first volume on matrix mechanics and then a second one on wave mechanics.[17] But most physicists had never bothered with matrix mechanics in the first place, and the few who had had already moved on to the much more fruitful version of quantum mechanics that incorporated waves. Perhaps, as Jordan had once said in dismissing Schrödinger's wave mechanics, he and Born really did believe that physics students should first be grounded in the Göttingen approach before being exposed to wave mechanics.

Pauli's review of the book was scathing. He roundly condemned Born and Jordan for leaving out wave mechanics, since many results in quantum physics could not be derived without them. He did have one positive thing to say: "The setup of the book as far as printing and paper are concerned is splendid."[18]

But then, Pauli didn't have a lot good to say about much of anything anymore. He was still in an angry, alcohol-fuelled fog. He confessed to a friend that he did not get on with women at all,[19] which was certainly reflected in his brief and misjudged marriage at the end of 1929 to a Berlin dancer.[20] It lasted only a few months before she left him for another man.

At the end of 1930, Pauli did something he'd never done before. A new field of research had developed around the nucleus of the atom, and experimental results seemed to indicate that the atom was not made up of just electrons and protons; something else was using up a tiny bit of energy every time a particle came out of the nucleus. Impulsively, and without thinking it through, Pauli declared that the energy was being used up by another very, very small particle that he called a neutron, although that was quickly changed to neutrino (little neutron).[21] It turned out to be a good guess, especially for someone who almost never guessed.

In 1931, Pauli spent three months lecturing in the United States. Prohibition was still in effect, which did not suit him at all. During a stay at the University of Michigan, he slipped across the border to Canada, where the liquor flowed (relatively) freely. He got drunk, fell and broke a bone in his shoulder. The official explanation for Pauli's continuing his lectures with a cast on was that he'd fallen getting off a boat.[22]

When Pauli started getting thrown out of bars in Zurich for his disruptive behaviour, his father finally stepped in. Pauli Sr. knew the Viennese psychoanalyst Carl Jung, and he advised his son to get some treatment, which prompted Pauli to read Jung's books and attend some of his lectures. In January 1932, Pauli finally talked to Jung after a lecture, and a week later Jung had set him up with one of his newly graduated assistants, Erna Rosenbaum. According to Pauli biographer Charles P. Enz, Jung had his own reasons for recommending Rosenbaum as his analyst: "since with Pauli's neurosis he was dealing with an underdeveloped feeling side, which appeared in the form of problems with women, a female analyst was indicated." Pauli was in analysis with the cheerful and bubbly Rosenbaum for eight months before he started working with Jung.[23]

Unlike Pauli, Schrödinger had never suffered from problems with women. Quite the opposite. He and Ithi had been lovers for several years. Anny was away for part of the summer in 1932, so Ithi was able to stay with Schrödinger at his house. She even got to meet

Planck and Einstein. Although Schrödinger had promised to take precautions to make sure she didn't get pregnant, it happened anyway. Schrödinger had always wanted children, so he encouraged Ithi to have the baby, but he did not offer to divorce Anny and marry her. Deeply upset, Ithi had an abortion, and that marked the end of their relationship.[24]

Ensconced in Leipzig, Heisenberg was not a particularly happy man either. The glow of what he considered his Solvay triumph had long since dissipated, and he was frustrated with his lack of progress in his own work on the atomic nucleus. In November 1930, Heisenberg's father died after contracting typhoid fever from the local drinking water at a conference in Greece.[25] The two men had never grown close, not even after Werner had achieved career success by becoming the youngest full professor in Germany. August Heisenberg's anxiousness about his son's ability to succeed, with its tacit anticipation of failure, seemed to have irreversibly tainted their relationship.

Heisenberg had been on a gruelling world tour that began in March 1929 and ended in November, and included India, Japan and the United States, spreading the Copenhagen gospel. Using his lecture notes from the tour, Heisenberg got to work on a book of his own, *The Physical Principles of the Quantum Theory*. For someone who was as ambitious as Heisenberg, the content of his book was decidedly odd. It was as if Bohr had discovered all of quantum physics by himself. Heisenberg even credited his own uncertainty relation to Bohr, without mentioning his own contribution. Indeed, the only citations in the book were for Bohr.[26]

If Heisenberg was still ashamed of how he'd treated Bohr during the spring and summer of 1927, he may have been going a little bit overboard to make amends, but in fact he and Bohr had mended fences back in 1928 when Heisenberg had made a return trip to Copenhagen. It is more likely that it suited Heisenberg to have Bohr, the philosopher, be seen as the authority on quantum theory.

There was no mistaking Bohr's belief—even pride—that his theory was philosophy and not physics. "I had to leave all physics out of my article," he wrote to Pauli in 1929, "and confine myself to pure philosophy . . ."[27] In his reply, Pauli congratulated him for his physics-free work but suggested he might want to leave the philosophy to the professionals.[28]

Heisenberg made a point of giving lectures to philosophers, including those of the Vienna Circle, encouraging them to look at the philosophical implications raised by such ideas as uncertainty and acausality. It wasn't, for instance, all that difficult to ascribe rules about non-visualizability and unknowability, along with a heavy reliance on abstract mathematics, to the influence of logical positivism, even if none of the creators of quantum physics were logical positivists. But such labels were being added after the fact to rationalize the Göttingen–Copenhagen version of quantum physics.

According to biographer David Cassidy, by 1931 Heisenberg had got philosophers involved in the quantum theory debate over meaning. And then he neatly separated out quantum mechanics as the purview of physicists and quantum theory as the realm of philosophers. Physicists, he said, had finished coming up with anything that could affect the philosophy of quantum physics years ago.[29] And as long as Heisenberg kept saying that the debates about quantum physics had all been settled at Solvay in 1927, the message to the rest of the physics community was that physicists such as Einstein, Schrödinger and de Broglie had nothing new to say.

It turned out that Louis de Broglie was not the failure his mother thought. The year after she died, believing her youngest son a "ne'er-do-well," he won the Nobel Prize for his theory of matter waves. The award, however, did not persuade him to revisit his pilot wave theory. De Broglie remained no less a target of Copenhagen ridicule than he had been before winning the Nobel Prize.

It's nothing new for "winners," particularly those who are self-declared winners, to conflate those who oppose the winner with the old order, and then dismiss critics as old-timers who can't get with· the new program or as people who aren't smart enough to keep up with the rapid advances. According to historian of science Mara Beller:

> In their fabricated narratives, the winners construct the profile of the opposition and the description of the past concurrently. The ideas of the opposition are projected as most characteristic of the overthrown past, and thus the opposition naturally appears reactionary—disposing of the old and discrediting the opposition are, in fact, one and the same process. As a result, not only is the opposition caricatured but past science is trivialized.[30]

Thus, Einstein could be caricatured as an old man whose best years were behind him because he disagreed with complementarity, whilst Bohr, who long since had given up on the heavily mathematical new physics, became a heroic figure. His disciples transformed what would have been unacceptable in any other physicist—his inability to do the necessary mathematics—into a virtue. "Bohr's devotees," says Beller, "believed that, unlike ordinary mortals, Bohr did not need to calculate in order to obtain 'the truth.'" Likewise, Bohr's unintelligibility as he struggled to explain his ideas was interpreted as a product of the great suffering he endured in his quest for understanding. "But the more incomprehensible Bohr's words became, the more his devotees' conviction grew that a great truth, inaccessible to simple mortals, was hidden therein."[31]

Bohr had certainly earned the respect and admiration of the many students and scientists who had passed through the doors of his Copenhagen institute. He'd earned his Nobel Prize, and developed his institute from a 10-by-15-foot office into a world-renowned centre. Bohr's institute was a place of great intellectual stimulation and simple fun, and he made a point of taking care of the people around him, both financially and as a father figure.

As a measure of the esteem in which Bohr was held by his Danish countrymen, he and his family were given the Residence of Honour in which to live, a sumptuous residence on the grounds of the Carlsberg Brewery. It came with nine acres of botanical gardens, a fountain in the house and lots of servants. The Bohr family moved into the grand house in the summer of 1932, except for their youngest son. Their sixth son, born in 1928, had been incapacitated as a baby by what may have been meningitis and remained in an institution.[32]

Bohr was not superhuman, yet his pronouncements on quantum theory were judged so subtle or so deep that even the world's leading physicists were not smart enough to fathom what he was saying. Such mythological powers were almost impossible to challenge. Bohr might well have been speaking gibberish, but who dared say so? Before long, physicists could not even question the shortcomings of quantum physics. Challenging quantum physics became conflated with doubting the great Bohr, and anyone who dared to criticize quantum physics quickly brought the wrath of the Bohr disciples upon his or her head.

Bohr and Einstein met again at the 1930 Solvay conference, and again they argued over the issue of non-separability. Bohr still felt that complementarity dealt with the problem, and Einstein still believed it merely hid the problem. But it was hard to argue about the success of quantum mechanics. The number of students studying it had mushroomed, and it was proving to be a very powerful tool for experimental discovery.

By the spring of 1933, Einstein had other problems to worry about. He was still in California when Hitler was appointed Chancellor of Germany in January. In March, Hitler seized control of the government. The Weimar democracy ceased to be, replaced by the Third Reich. Hitler suspended political and civil rights, and the National Socialist Party, which had effectively already taken over the streets of the cities, now took control of the entire country.[33] It was clear to Einstein that he could not return to Germany as long as the current regime was in power. He was not prepared, he said publicly, to live in a country without civil liberties and freedom of speech. Einstein's celebrity status ensured that his public condemnation of Hitler's government quickly reached the ears of those in power. The government froze Einstein's bank accounts and police raided his apartment in Berlin. Fortunately, his step-daughter Margot had already packaged up most of Einstein's important documents and sent them to the French embassy.[34] Instead of returning to Germany in the spring as scheduled, Albert and Elsa settled in Belgium.

When the Nazis came to power, the university student governments wanted to see immediate action. Since university professors were government employees, it was a simple matter to take action against them. On April 7, 1933, the new Law for the Restoration of the Professional Civil Service came into effect; it removed "non-Aryans" and "politically unreliable" people from their civil service positions.[35] The effect on universities was immediate and dramatic. The Göttingen student union had been working on a list of Jewish professors, and it didn't take long for the purge to begin. James Franck resigned in protest against the law as soon as it was announced. Three weeks later, Max Born and Richard Courant were suspended with pay, along with four other Göttingen professors. All of them were forbidden from setting foot in a university classroom. Born was devastated. "All I had built up in Göttingen, during twelve years' hard work, was shattered. It seemed to me like the end of the world."[36]

Pascual Jordan, who had worked with both Born and Franck, went to see them at Göttingen. He was very upset about Born's suspension and Franck's resignation. After the trip, he wrote to his mother that he might have been able to do something to help them if he had belonged to the Nazi Party.[37] Soon after, he became a Nazi storm trooper.

Jordan already had strong feelings about German nationalism. He'd been writing essays for the nationalist journal *Deutsches Volks-*

tum since 1930 under the pseudonym Ernst Domeier.[38] His essays railed against liberalism and the negative effects of Marxism on society, and espoused the value of science and technology for military power. In 1933, Jordan was appointed to a professorial position at a minor German university, so he no longer had to hide behind a pseudonym for fear of damaging his career prospects. As a professor and a member of the ruling party, he could now write his essays under his own name.

Hedi Born, who had rallied in the face of crisis, started organizing the Born family's departure from Germany, not knowing if or when they might return. They left Germany in May, leaving behind almost all their possessions.

Erwin and Anny Schrödinger were on the move as well in the summer of 1933. Schrödinger was not Jewish, but he had no desire to remain in Berlin under the Nazi regime. He was careful not to make any public statements about his feelings, but he was already considered one of the "politically unreliable." Schrödinger quietly arranged for positions at Oxford University for himself and his assistant, Arthur March, since Erwin had fallen in love with Arthur's wife, Hilde.

The Schrödingers and Marches spent part of the summer in northern Italy, near the same resort community where Max and Hedi Born were staying while they all tried to figure out where to go next. Hermann Weyl, Anny's lover, had resigned from Göttingen University and was there as well.[39]

Pauli, who was nearby in Zurich, and Max each wrote to Werner Heisenberg, but both were unsuccessful in persuading him to join them in the mountains. Heisenberg was expecting them all back for the fall term, and he advised Born to return to Göttingen and just wait for the whole suspension thing to sort itself out. Born was furious at how dismissive and lacking in empathy Heisenberg was about the upheavals he was facing, and wrote him a long and emotional letter describing how it felt to be suspended and to lose his home.[40]

The future was unpredictable for them all, but for a while the Schrödingers, Borns and Marches could enjoy the beautiful mountain scenery and the company of friends. Even Max and Hedi were getting along much better. It also wasn't long before Hilde found out she was pregnant. The Marches had been married for four years and the Schrödingers for thirteen, and neither couple had children.[41] The prospect of a baby suited them all. They could all be parents together for Erwin and Hilde's child.

Schrödinger wasn't doing much thinking about physics with so many other things on his mind, but he took time to write to Einstein, hoping to see him at the upcoming Solvay conference in October. "Unfortunately," he said, "(like most of us) I have not had enough nervous peace in recent months to work on anything seriously."[42]

Albert and Elsa Einstein spent the summer in a seaside resort community in Belgium. The Belgian government and the royal family were concerned for Einstein's safety, since there were rumours of a possible assassination attempt. Einstein was assigned two bodyguards, but he did not much care for being under police supervision. Inhabitants of the town were told not to answer anyone who asked where Einstein lived. Philipp Frank was passing near the town and decided to stop in for a visit with his long-time friend, unaware of the precautions that had been made for Einstein's protection. Frank had no difficulty at all getting directions to the villa where Albert and Elsa were living. As he approached it, however, he was seized by the bodyguards, who had mistaken him for a potential assassin. Einstein, who had lived with death threats before, thought it all very funny.[43]

But there was nothing funny about what was happening to German universities under the Nazi regime. Hitler wanted science to be an instrument for advancing national pride, but he wanted it to be German science. Anything the Jews such as Einstein had worked on was tainted, which meant that Hitler would not support continuing research in relativity or quantum physics. Even the mathematics department at Göttingen had been gutted, with Gustav Herglotz the sole remaining full-time professor. Max Planck, deeply concerned about the loss of so many superb scientists wrought by the Restoration Law, went to see Hitler to plead for exemptions for Jewish scientists. Hitler wasn't interested. "If science cannot do without Jews," he told Planck, "then we will do without science for a few years."[44]

In just a few short months, Sommerfeld's "nursery" for quantum physicists in Munich and Born's mathematical physics dynasty in Göttingen were gone. The buildings were still standing, but the environment that had nurtured the struggle to understand the quantum world simply ceased to be. Of the three schools that had developed so much of quantum physics, only Bohr's temple in Copenhagen remained.

As more and more scientists lost their positions at German universities, Ehrenfest was feeling increasingly guilty about continuing to occupy one of the prime chairs in Europe. It didn't help that Max Born was getting offers from universities in Paris, Leningrad, Brussels, Belgrade and Constantinople[45] while Ehrenfest received nothing. Of course, their circumstances were quite different: Born was being forced out of his job; Ehrenfest wasn't. There was a depression going on, and with so many German scientists trying to find jobs outside the country, job opportunities were scarce for most of them.

In June, Ehrenfest wrote another plaintive letter about his misfortunes to his brother Hugo. However, it seemed that Hugo was getting thoroughly fed up with his youngest brother. He wrote back forcefully:

> To hell with your damned crankiness. You use your large intelligence wastefully and readily for the solution of the most difficult problems of others. Rarely or never for yourself. Personally I am convinced that not only is my theory generally correct, but it also is particularly true regarding your own

scientific activity. And here, perhaps, lies the source of your depression.

Hugo had said all this before, yet it seemed to do nothing to help Paul. "But what's the use," Hugo ended his letter. "For God's sake, don't be a crybaby!"[46]

Ehrenfest spent some of the summer with Nell at a hotel on an island near the North Sea,[47] but he still couldn't seem to find any peace.

Born finally got the job offer he'd been waiting for in July. With a little help from Einstein, who was in England at the time, Ernest Rutherford offered Born a position at Cambridge. He didn't accept the job immediately, though. He wrote to Ehrenfest that he too was feeling guilty, because he feared he was taking a position away from a deserving English scientist. He also wrote to Ralph Fowler about his concern, but Fowler assured him that the position had been created especially for him and would not be offered to anyone else if Born didn't take it.[48] So he accepted the job, greatly relieved that he now knew where he and his family would be at the end of the summer.

By the fall, Tatyana Ehrenfest had returned to Moscow to teach,[49] Erwin and Anny Schrödinger were settled in Oxford, and Max and Hedi Born and their children were in Cambridge, where Dirac was a professor. Dirac had been getting job offers too. He'd been offered a position at the University of Toronto the previous year, but had turned it down. Dirac had had his eye on the prestigious Lucasian chair of mathematics at Cambridge, a position once held by Isaac Newton, and his wish was fulfilled in the fall of 1932.[50]

In September, Dirac made a point of going to Copenhagen for one of the regular meetings Bohr organized. In 1929, Bohr had invited a number of scientists to an informal week-long gathering at Copenhagen. The event had no official agenda, so they could talk about whatever they liked. It had gone so well that Bohr started holding the meetings every year on the Easter weekend. In 1933, however, Bohr had been in the United States in the spring, so the gathering was being held in the fall instead, from September 13 to 20.[51]

Ehrenfest was there too. He participated in the discussions, but sometimes he seemed agitated. Dirac noticed it, especially when they ran into each other at the Bohrs' big house at the end of the get-together. Dirac complimented him on his contributions to the discussions, which seemed to fluster Ehrenfest. "What you have said," Ehrenfest told him with great emotion, "coming from a young man like you, means very much to me, because, maybe, a man such as I feels he has no longer the force to live."[52] Dirac had never been comfortable with emotional outpourings, and he was rather alarmed—with good reason, as it turned out.

On September 25, Ehrenfest headed for the train station to go to the institute in Amsterdam where Vassily was being cared for. He was carrying a handgun, which he may well have learned how to use while serving in the home guard, as many men had done during the Great War. On the way he stopped in to see a psychiatrist he was acquainted with at his office, not far from the train station. Ehrenfest and the psychiatrist had what appeared to be a casual chat, and the psychiatrist didn't sense anything wrong; to him, Ehrenfest was his normal self.[53] Ehrenfest may have seemed quite calm, as is sometimes the case with people who have finally dealt with intolerable emotional turmoil by making the decision to commit suicide.

In the waiting room of the institute where he met with Vassily, Ehrenfest shakily pulled out the gun and shot his son. Ehrenfest was most likely unaware that Vassily was not killed outright before he turned the gun on himself. The boy, however, did not survive long. The Dutch newspapers the next day announced the deaths of both father and son.[54]

Einstein was in England when the news of his friend's suicide reached him. There was nothing he could do but mourn the loss of yet another friend, and of yet another of the happy families where he had once found such warmth and companionship. In the memoriam Einstein wrote for Ehrenfest, he didn't gloss over his friend's emotional difficulties, which, he said,

> were aggravated by the strangely turbulent development which theoretical physics had recently undergone. To learn and to teach things that one cannot fully accept with one's heart is always a difficult matter, doubly difficult for a mind of fanatical honesty, a mind to which clarity means everything.

Einstein also talked about Tatyana.

> The strongest relationship in his life was that to his wife and fellow worker, an unusually strong and steadfast personality and his intellectual equal . . . Her steadfastness in the face of hardships, her integrity in thought, feeling and action—all these were a blessing to him and he repaid her with veneration and love such as I have not often witnessed in my life.[55]

On October 3, Einstein gave a lecture to a packed house at the Royal Albert Hall. As he ended his speech, he praised the solitude of the scholar who needs to be alone to think through scientific problems. He received a standing ovation. For all his celebrity, according to biographer Jürgen Neffe, he was a shockingly lonely man.[56]

The Seventh Solvay Conference opened on October 20, 1933. Paul Langevin, who had replaced Lorentz as conference chair, took time at the beginning of the gathering to pay tribute to Ehrenfest. Langevin described him as having been "at the very heart of the drama of contemporary physics," and said that he had personified that drama in his own life.[57]

Ehrenfest's life did, very much, mirror the development of quantum physics. He had been with Boltzmann, who started it all with his imaginary "chunking" of energy, and worked with both Planck's and Einstein's ideas on quantization. He too had struggled to understand the baffling quantum world, witnessing the tension between Bohr and Einstein, his two dear friends, as they sought to understand as well. Bohr had arrived at a theory that worked for him, but he was never really satisfied with it as long as he couldn't persuade Einstein to see his point of view. Einstein refused to accept a theory that he believed merely hid the problem. Bohr struggled with it. Ehrenfest, finally, surrendered.

Chapter Eighteen

QUANTUM CONFUSION

The quantum theory has been very like other victories:
you smile for months and then weep for years.

HENDRIK KRAMERS

"Western philosophy is often said, in only part exaggeration, to be a long footnote to Plato," said Harvard historian of science Peter Galison at the beginning of the Twenty-third Solvay Conference in 2005. "The physics of this last century, documented—and now reawakened—in the *Conseils de Solvay*, might be similarly seen as a long elaboration of the great dialogue Einstein initiated between relativity and the quantum."[1] The struggle by physicists to make sense of the quantum, which began with the First Solvay Conference in 1911 and reached a turning point at the meeting in 1927, was picked up once again at the Solvay meeting in 2005 on string theory (which should more accurately be called a hypothesis or proposal, since it remains untested).

Einstein, Ehrenfest and Oskar Klein had toyed with higher dimensions in the summer of 1926 as a means for unifying relativity and the quantum, and with it they sowed the seeds for string theory. They weren't the only physicists who recognized that, theoretically speaking, Bohr's quantized atom in three spatial dimensions appeared to become a classical atom in four. German physicist Theodor Kaluza had been the first to suggest adding the fourth spatial dimension, back in 1919, but it was Klein who proposed the idea that the extra dimension couldn't be observed in the three-dimensional world because, although it was everywhere in space, it was rolled up in a whole lot of very, very tiny circles, each smaller than a Planck length. For a short while in 1926, higher dimensions had seemed to hold the key to unifying relativity and the quantum. The Kaluza–Klein universe, resurrected in the 1970s and then again in the 1980s, was the jumping-off point for a string theory that would, by the 2005 Solvay conference, be sporting nine or ten dimensions.

The conference in Brussels in December 2005 was only a three-day affair, similar in structure to the 1927 conference but without the two days of general discussion once the formal sessions were over. As for previous conferences, the invited guests stayed in the luxurious Hôtel Métropole. But this time the sessions all took place in the hotel, in the Excelsior Room, with marble flooring, panelled walls, twelve-foot ceilings and large chandeliers, which typically was rented out for 2,500 a day. It was a suitably grand venue for the scientific elite.*

The theme of the 2005 conference was "The Quantum Structure of Space and Time," focusing primarily on string theory. The meeting was chaired by physicist and Nobel laureate David Gross, a leader in string research. Besides Gross, there were four other Nobel Prize recipients invited: Murray Gell-Mann, Steven Weinberg, Gerard 't Hooft and Frank Wilczek, along with a number of prizewinning mathematicians. Of the sixty invited scientists, more than half were from American universities or institutions, reflecting the seismic shift from Germany as the centre of physics before the Second World War to the United States afterwards. There had been no American physicists at all at the first Solvay meeting in

---

*The 2005 Solvay conference was the first time participation by the general public was allowed, with more than nine hundred people registering to attend a free Sunday afternoon session featuring talks by such notables as string theorist and author Brian Greene and Nobel laureates David Gross and Gerard 't Hooft.

1911 on "Radiation Theory and the Quanta," and only two out of about thirty scientists at the 1927 conference on "Electrons and Photons."* And, unlike the 1911 and 1927 Solvay conferences, where Marie Curie was the lone female scientist, the 2005 conference boasted three.†

Much like the 1911 and 1927 conferences, however, the 2005 event ended in confusion. Gross, in his wrap-up speech at the end of the conference, admitted that scientists were in a "period of utter confusion."[2] In fact, he noted that physicists today are in a similar conundrum with quantum theory as those who were at the 1911 meeting.

After the First Solvay Conference, Einstein wrote to his long-time friend Michele Besso about the debates at the event, telling him that "one lamented at the failure of the theory without finding a remedy. This Congress had the aspect similar to the wailing at the ruins in Jerusalem. Nothing positive came out of it."[3] Similarly, Paul Langevin declared that the Fifth Solvay in 1927, with its presentation of three different versions of quantum physics, was the ultimate "confusion of ideas."[4]

The history of the development of quantum physics has been punctuated by difficult periods where its creators were groping in the dark, trying to figure out some way to make sense of the quantum world. Not much has changed. In the fall of 2007, Gross reiterated his strong support for string theory as the most likely source of a unified theory, without being blind to the fact that string theorists are routinely working in the dark as well. "Even those of us who work in the field aren't really sure what string theory is or what it's

---

*Light quanta were called photons after 1926.

†The three female scientists at the 2005 Solvay conference were Lisa Randall, professor of physics at Harvard University and author of the 2006 book on strings, gravity and higher dimensions *Warped Passages*; Renata Kallosh, professor of physics at Stanford University; and Eva Silverstein, an associate professor of physics, also at Stanford.

going to be. When you're in this kind of speculative, exploratory science, it's important to have faith because you're out on a limb."[5]

As Galison noted at the start of this chapter, the great dialogue that began nearly a century ago has been reawakened, although it's never really gone away. The remarkable experimental success of quantum mechanics—which has brought the world everything from nuclear energy and the atomic bomb to CD players and computers—has tended to overshadow the continuing failure of physicists to put the universe under a single set of physics rules.

Perhaps the discomfort physicists are feeling, as it becomes more difficult to sustain faith that string theory is the keeper of the Holy Grail of unification, might make it easier to overcome the entrenched resistance in the discipline to looking that old quantum elephant in the eye. If the reason that the very smart and talented people working in physics cannot reconcile quantum physics with relativity is that quantum physics is somehow flawed or mistaken, it's time to open it up to discussion again. If it's relativity that's the problem, well, that's another story.

Nearly a century after Paul and Tatyana Ehrenfest completed their book-length encyclopedia article in St. Petersburg entitled "The Conceptual Foundations of the Statistical Approach in Mechanics," it remains a classical text in statistical mechanics, and is still in print. The most recent hardcover English edition was published in 2002 by Dover Publications.

Wolfgang Pauli's 1921 encyclopedia article *Theory of General Relativity*, written while he was doing his doctoral degree in Munich, is also still available from Dover.

Paul Ehrenfest Jr. grew up to become a physicist just like his father. Young Ehrenfest was part of a team, headed by French scientist Pierre Auger, that was studying the strange phenomenon of cosmic rays. In the 1930s the mysterious extraterrestrial rays were causing considerable excitement among scientists. Laboratory equipment wasn't powerful enough to test the high-energy rays, so physicists were attempting to track them in deep mines, on the ocean or in the air via balloons and airplanes.

In 1938, Auger, Ehrenfest and the rest of the team hauled their experimental equipment up into the Alps to a height of 3,500 feet above sea level. They were the first to identify high-energy showers of particles as coming from cosmic rays, and published their results in the *Review of Modern Physics* in the summer of 1939.

Ehrenfest's career, however, was very short-lived. He was killed in a mountain avalanche while skiing in the Alps that same year, at the age of twenty-four.[1]

One of the factors that led to the suicide of Paul Ehrenfest was his belief that one of the many German physicists who had been fired by the Hitler regime should have his job, but after his death that's not what happened. Instead, Ehrenfest's position as chair of physics at Leiden went to his former student Hans Kramers, who had become professor of theoretical physics at the Dutch University of Utrecht after serving as Niels Bohr's first assistant in Copenhagen.

In turn, Kramers's chair in Utrecht was filled by another Ehrenfest student, George Uhlenbeck, who returned to Holland from the United States to take up the position in 1935. Neither man was Jewish.

The guest bedroom in the Ehrenfest home at 57 Witterozenstraat in Leiden was a temporary resting place for many scientists. Paul Ehrenfest had his guests and their wives sign their names on a wall behind a tapestry in the bedroom. Of the many signatures hidden behind the wall hanging, fourteen belonged to men who were—or would become—Nobel laureates.

Ehrenfest also had speakers he invited to his Wednesday evening colloquia at the university sign a wall in the theoretical physics department. Other visiting scientists and mathematicians would sign the wall as well. When the old building was torn down in 1962, the part of the wall with the signatures was preserved, and then installed in the new colloquium room, where it remains today.*

The Nobel committee for physics had a very difficult time assessing who should receive a prize for their contributions to the devel-

---

*A picture of the signature wall is available online at http://ilorentz.org/history/colloquium/muur_heel.html.

opment of quantum physics, partly because the new science was still so new and confusing—and partly because most of the committee members worked in fields unrelated to the quantum and didn't understand it very well. The 1929 prize went to Louis de Broglie, but the bitter controversy between the committee members over which of the quantum founders should get the next award resulted in it finally being given to Indian physicist Chandrasekhara Vankata Raman for his work on light scattering, which was unrelated to the controversial quantum.

No prize in physics was awarded in 1931, again because the committee couldn't agree on who should receive it and because some members believed it was premature to be giving prizes for the new physics. In 1932 the acrimonious debate on the Nobel committee continued, and awarding of the prize was put off for another year.

With considerable pressure being placed on the Nobel committee by people such as Albert Einstein, Max Planck and Niels Bohr, it finally conceded in 1933 that quantum physics was proving its worth. Werner Heisenberg got the nod for the delayed 1932 award, with Paul Dirac and Erwin Schrödinger jointly awarded the 1933 prize.

The rest of the quantum founders had to wait awhile for Nobel recognition. Wolfgang Pauli got the award in 1945, but Max Born did not receive his until 1954. Born had received many nominations, but the Nobel committee had reservations about giving Born an award for work he'd done with Pascual Jordan because of Jordan's association with the Nazi regime. Born's award, in the end, was for his work on probabilities, which he'd done on his own. Jordan was repeatedly nominated for a Nobel Prize but never received one.

Max and Hedi Born's children grew up in England, where Max had a position at Cambridge University. Their older daughter, Irene, married a Welsh professor, and they had three children. The youngest child of Irene and Brinley Newton-John—Olivia—was born in 1948 and went on to become a pop singer.

CHAPTER ONE

1 Boutin, Paul, 2005.
2 Greene, Brian, *The Elegant Universe*, Vintage Books, 2000, p. 210.
3 Hooft, Gerard 't, 2001.
4 Weinberg, Steven, 2003.
5 Plato, as quoted by Philip E.B. Jourdain in "The Nature of Mathematics," *The World of Mathematics*, Vol. 1, ed. James R. Newman, Simon and Shuster, 1956, p. 15.
6 Gershenson, Daniel E., and Daniel A. Greenberg, 1964.
7 See, for instance, Lee Smolin's *The Trouble with Physics*, 2006, and Peter Woit's *Not Even Wrong*, 2006.
8 Murray Gell-Mann, as quoted by Wolpert, L., 1993, p. 144.
9 Weinberg, Steven, 1992, p. 84.
10 Johnson, George, 2007.

CHAPTER TWO

1 Paul Ehrenfest, in a letter to Albert Einstein, August 16, 1920, as quoted by Klein, Martin J., 1970, p. 319.
2 Wolfgang Pauli, in a letter to Gregor Wentzel, December 5, 1926, as quoted by Segrè, Gino, 2007, p. 81.
3 Peat, F. David, 1987, p. 15.
4 Segrè, Gino, 2007, p. 79.
5 Quinn, Susan, 1995, pp. 309–25.
6 Albert Einstein, as quoted by Mehra, Jagdish, 1975, p. xiv.
7 Klein, Martin J., 1970a, p. 175.
8 Moore, Walter, 1989, p. 239.
9 Moore, Walter, 1989, p. 175.
10 H.A. Lorentz, in a private note on the back of a telegram, in the collection of the Solvay Institutes, Brussels, as quoted by Mehra, 1975, p. xxiv.
11 Albert Einstein, as quoted by Levenson, Thomas, 2003, p. 270.
12 Moore, Ruth, 1966, p. 163.
13 Harald Bohr, as quoted by Klein, Martin J., 1981.

CHAPTER THREE

1   Kragh, Helge, 1999, p. 20.

2   The Department of Mathematics at the North Dakota State University runs the Mathematics Genealogy Project, available online at http://www.genealogy.ams.org. According to the MGP, Joseph Lagrange had three students. One student, Jean-Baptiste Fourier, has more than 22,750 descendants who went on to become mathematicians or physicists. Another Lagrange student, Siméon Poisson, has more than 33,000 descendants. This genealogy traces the "descendants" of a mathematician/physicist from the students he trained or supervised (we'd call them graduate or postgraduate students today) to the students they supervised, and so on, to the present day.

3   Ebeling, Werner, and Dieter Hoffman, 1991.

4   Reiter, Wolfgang L., 2007, pp. 361–62.

5   Blackmore, John, 1985.

6   Klein, Martin J., 1964b, p. 84.

7   Max Planck was the first to write the entropy equation as $S = k \log W$, where S is entropy, k is Boltzmann's constant, and $\log W$, in the natural logarithm of the probability W. The equation appears in this form on Boltzmann's tombstone in the Vienna Central Cemetery.

8   Max Planck, as quoted in Klein, Martin J., 1964b, p. 105.

9   Ludwig Boltzmann, as quoted by Ostwald, Wilhelm, 1927.

10 Historians of science such as Thomas Kuhn, Martin J. Klein, Olivier Darrigol, Clayton Gearhart and Allan Needell, in particular, have spent considerable time analyzing this important time in history, but without arriving at any consensus on the rationale behind Planck's quantization.

11 Klein, Étienne, 1997, p. 50.

12 Reiter, Wolfgang L., 2007, p. 365.

13 Ibid.

14 Wilhelm Ostwald, as quoted by Klein, Martin J., 1970a, p. 76.

15 Ibid., p. 42.

16 Ibid., p. 46.

17 Ibid., p. 18.

18 Ogilvie, Marilyn Bailey, and Joy Dorothy Harvey, eds., 2000, p. 407.

19 There is no reference to Tatyana's mother being part of her life while she was growing up, but she did live with Paul and Tatyana at times in St. Petersburg and Leiden, helping care for the Ehrenfest children.

20 Klein, Martin J., 1970a, p. 42.

21 Ibid., p. 50.

22 Reiter, Wolfgang L., 2007, p. 369.

23 Wilhelm Ostwald, as quoted in Klein, Martin J., 1970a, p. 77.

24 Huijnen, Pim, and A.J. Knox, 2007, p. 188.

25 Klein, Martin J., 1970a, p. 121.

26 Huijnen, Pim, and A.J. Knox, 2007, p. 191.

27 Ibid., p. 191.

28 Ibid., p. 195.

29 Klein, Martin J., 1970a, p. 94, with an excerpt from Albert Einstein, "Paul Ehrenfest in Memoriam," *Out of My Later Years*, Philosophical Library, 1950, p. 215.

30 Blackmore, John, 1985, p. 301.

31 Klein, Martin J., 1970a, p. 121.

CHAPTER FOUR

1 Klein, Martin J., 1970a, p. 5.
2 Paul Ehrenfest, in a letter to Abram Joffe, February 20, 1913, as quoted in Klein, Martin J., 1970a, p. 15.
3 Klein, Martin A., 1970a, p. 176.
4 Highfield, Roger, and Paul Carter, 1993, p. 19.
5 Ibid., pp. 29–31.
6 Albert Einstein never completely lost his connection to the Winteler family. His sister Maja married Marie's brother Paul, and Albert's best friend, Michele Besso, married Marie's sister Anna.
7 Highfield, Roger, and Paul Carter, 1993, p. 31.
8 Ibid., p. 46.
9 Pais, Abraham, 1982, p. 44.
10 Ibid., p. 44.
11 Ibid., p. 46.
12 Mehra, Jagdish, 1975, p. xx.
13 Albert Einstein, as quoted by Hoffman, Banesh, with Helen Dukus, 1972, p. 36.
14 Albert Einstein, in a letter to Michele Besso, January 22, 1903, in Einstein, Albert, 1995, p.7.
15 Klein, Martin J., and Allan Needell, 1977, p. 601.
16 Rigden, John S., 2005, p. 31.
17 Ibid., p. 11.
18 Klein, Martin J., 1964a, p. 66.
19 Albert Einstein, as quoted by Rigden, John S., 2005, p. 32.
20 Stuewer, Roger H., 2006, pp. 544–55.
21 Klein, Martin J., 1964a, p. 60.
22 Ibid., p. 61.
23 Ibid., p. 81.
24 Highfield, Roger, and Paul Carter, 1993, pp. 23–24.
25 Klein, Martin J., 1970a, p. 217.
26 Ibid., p. 245.
27 Ibid., pp. 252–53.
28 Ibid., p. 253.

CHAPTER FIVE

1 Klein, Martin J., 1970a, p. 204.
2 J.M. Burgers, from his unpublished *Autobiographical Notes*, as quoted in Klein, Martin J., 1970a, pp. 204–5.
3 Martin J. Klein, personal communication.
4 van Lunteren, Frans, 2001.
5 Klein, Martin J., 1970a, p. 311.
6 Albert Einstein, in a letter to Paul Ehrenfest, November 9, 1919, as quoted in Klein, Martin J., 1970a, pp. 313, 314.
7 Moore, Ruth, 1966, p. 6.
8 Ibid., p. 6.
9 Ibid., p. 21.
10 Pais, Abraham, 1991, p. 101.

11 Ibid., p. 107.

12 Ibid., p. 108.

13 Ibid., p. 112.

14 As quoted in Moore, Ruth, 1966, p. 29.

15 Pais, Abraham, 1991, p. 128.

16 Niels Bohr, in a letter to Harald Bohr, June 12, 1912, as quoted by Beller, Mara, 1999a, p. 261.

17 Niels Bohr, in a letter to Harald Bohr, June 19, 1912, as quoted by Heilbron, John L., and Thomas S. Kuhn, 1969, pp. 238–39.

18 Niels Bohr, in the Rutherford memorandum, as quoted by Heilbron, John L., and Thomas S. Kuhn, 1969, p. 248.

19 E.N. da C. Andrade, in *The Collected Papers of Rutherford*, Vol. 2, *Interscience*, 1963, p. 299, as quoted in Pais, Abraham, 1991, p. 129.

20 Heilbron, John L., and Thomas S. Kuhn, 1969, p. 256.

21 Ibid., p. 258.

22 Curtis, Theodor, 1961, p. 579.

23 Ernest Rutherford, in a letter to Niels Bohr, March 20, 1913, as quoted by Pais, Abraham, 1991, p. 153.

24 Robert Pohl, recalling a comment by Emil Warburg, as quoted by Pais, Abraham, 1991, p. 154.

25 Paul Ehrenfest, in a letter to Hendrik Lorentz, August 25, 1913, as quoted in Klein, Martin J., 1970a, p. 278.

26 Pais, Abraham, 1991, p. 192.

27 Klein, Martin J., 1986, pp. 328–29.

28 Albert Einstein, in a letter to Paul Ehrenfest, March 22, 1919, tr. Bertram Schwarzschild, *Physics Today*, 58, No. 4, p. 88.

29 Niels Bohr, in a letter to Paul Ehrenfest, May 10, 1919, as quoted by Klein, Martin J., 1986, p. 330.

30 Albert Einstein, in a letter to Paul Ehrenfest, November 9, 1919, as quoted by Klein, Martin J., 1986, p. 331.

CHAPTER SIX

1 The Lorentz Medal has been awarded to twenty physicists since the first award in 1927 to Max Planck, but no female physicist has yet made the list.

2 Klein, Martin J., 1986, p. 336.

3 Pais, Abraham, 1991, p. 170.

4 Moore, Ruth, 1966, p. 99.

5 Niels Bohr, in a letter to O.W. Richardson, August 15, 1918, as quoted by Pais, Abraham, 1991, p. 192.

6 Ibid., p. 171.

7 Albert Einstein, in a letter to Michele Besso, May 13, 1911, as quoted by Pais, Abraham, 1982, p. 405.

8 Albert Einstein, as quoted by Pais, Abraham, 1991, p. 347.

9 Albert Einstein, in a letter to Michele Besso, July 29, 1918, as quoted by Pais, Abraham, 1982, p. 411.

10 Albert Einstein, in Langevin, Paul, and Maurice de Broglie, eds., 1912, pp. 429, 436.

11 Niels Bohr, 1923, as quoted by Pais, Abraham, 1991, p. 196.

12 Klein, Martin J., 1986, p. 331.

13 Niels Bohr, in a letter to Albert Einstein, June 24, 1920, as quoted in Klein, Martin J., 1986, p. 331.

14 Dresden, Max, 1987.

15 Ibid., p. 486.

16 Ibid., chap. 14. Note that "light quanta" has been substituted for "photons," a term not used in physics until 1926.

17 Niels Bohr, in a letter to Paul Ehrenfest, [1921?], as quoted by Boya, Luis J., 2003, p. 2566.

18 Klein, Martin J., 1970b, p. 23.

19 Pais, Abraham, 1991, p. 234.

20 Hendrik Kramers, as quoted by Boya, Luis J., 2003, p. 2566.

21 Niels Bohr, as quoted by Pais, Abraham, 1991, p. 229.

22 John Slater, as quoted by Boya, Luis J., 2003, p. 2567.

23 John Slater, in a letter to his family, January 18, 1924, as quoted by Pais, Abraham, 1991, p. 235.

24 Klein, Martin J., 1970b, p. 25.

25 J.C. Slater, clarifying the BKS paper in December 1924, *Physics Review*, 25, p. 395, as quoted by Pais, Abraham, 1991, p. 236.

26 Klein, Martin J., 1986, p. 333.

27 Klein, Martin J., 1970b, p. 28.

28 Max Born, in a letter to Niels Bohr, March 1924, as quoted by Thorndike Greenspan, Nancy, 2005, p. 121.

29 Albert Einstein, in an article on the Compton experiment for *Berliner Tageblatt*, April 20, 1924, as quoted by Klein, Martin J., 1970b, p. 38.

30 Franz Haber, in a letter to Albert Einstein, undated, 1924, as quoted by Pais, Abraham, 1991, p. 237.

31 Klein, Martin J., 1970b, p. 34.

32 Ibid.

33 Wolfgang Pauli, in a letter to Hendrik Kramers, July 27, 1925, as quoted by Pais, Abraham, 1991, p. 238.

34 Paul Ehrenfest, in a letter to Niels Bohr, September 19, 1925, as quoted by Klein, Martin J., 1986, p. 337.

CHAPTER SEVEN

1 Max Born, in a letter to Albert Einstein, August 25, 1923, in Einstein, Albert, and Max Born, 1971, p. 81.

2 Thorndike Greenspan, Nancy, 2005, p. 155.

3 Hedi Born, in a letter to Elsa Einstein, November 18, 1920, as quoted by Thorndike Greenspan, Nancy, 2005, p. 101.

4 Albert Einstein, in a letter to Max Born, January 30, 1921, as quoted by Thorndike Greenspan, Nancy, 2005, p. 101.

5 Max Born, in a letter to Albert Einstein, February 21, 1921, in Einstein, Albert, and Max Born, 1971, p. 51.

6 Thorndike Greenspan, Nancy, 2005, p. 29.

7 Ibid., p. 31.

8 Ibid., pp. 35–36.

9 Ibid., p. 43.

10 Born, Max, 1969, p. 107.

11 Ibid.

12 Thorndike Greenspan, Nancy, 2005, pp. 60–62.

13 Born, Max, 1969, p. 107.

14 Albert Einstein, in a letter to Hedi Born, February 8, 1918, in Einstein, Albert, and Max Born, 1971, p. 5.

15 Albert Einstein, in a letter to Max Born, January 27, 1920, in Einstein, Albert, and Max Born, 1971, p. 21.

16 Max Born, in a letter to Elsa Einstein, June 21, 1920, in Einstein, Albert, and Max Born, 1971, p. 30.

17 Max Born, in a letter to Albert Einstein, February 12, 1921, in Einstein, Albert, and Max Born, 1971, p. 54.

18 Eckert, M., 2000.

19 Max Born, in a letter to Wolfgang Pauli, December 23, 1919, as quoted by Thorndike Greenspan, Nancy, 2005, p. 109.

20 Max Born, in a letter to Albert Einstein, February 12, 1921, in Einstein, Albert, and Max Born, 1971, pp. 53–54.

CHAPTER EIGHT

1 Max Born, as quoted by Thorndike Greenspan, Nancy, 2005, p. 111.

2 Einstein, Albert, and Max Born, 1971, p. 63.

3 Peierls, R.E., 1960, p. 176.

4 Wolfgang Pauli, as quoted by Peierls, R.E., 1960, p. 186.

5 Werner Heisenberg, recalling Pauli's criticisms when they were students at Munich, as quoted by Cassidy, David C., 1992, p. 109.

6 Smutný, František, 1990.

7 Comte, Auguste, 1842.

8 Mach, Ernst, 1886.

9 Holton, Gerald, 2003.

10 Frank, Philipp, 1941, p. 6.

11 Brush, Stephen G., 1980, p. 407.

12 Ibid., pp. 407–8.

13 Stephen G. Brush, as quoted by Badash, Lawrence, 1972, p. 49.

14 Poincaré, Henri, 1905.

15 Albert Einstein, in his 1916 obituary for Ernst Mach, as quoted by Holton, Gerald, 1992, p. 27.

16 Smutný, František, 1990, p. 259.

17 Brush, Stephen G., 1980, p. 409.

18 See, for instance, a letter from American philosopher William James to his wife after meeting Ernst Mach in Prague in 1882, where James wrote of Mach: "I don't think anyone ever gave me so strong an impression of pure intelligent genius. He apparently read everything and thought of everything, and has an absolute simplicity of manner and winningness of smile, when his face lights up, that are charming." From James, William, 1926.

19 Wolfgang Pauli, in a letter to Carl Jung, March 31, 1953, as quoted by Meyenn, Karl von, and Engelbert Schucking, 2001.

20 Klein, Martin J., 1970a, p. 35.

21 Meyenn, Karl von, and Engelbert Schucking, 2001.

22 Jacobi, Manfred, 2000. See also Seppi, Ruth, 2004.

23 Peierls, R.E., 1960, p. 185.

24 Pais, Abraham, 1991, p. 201.

25 Ibid.

26 Paul Ehrenfest and Wolfgang Pauli, as quoted by Cline, Barbara Lovett, 1965, p. 139.

27 Cassidy, David C., 1992, pp. 192, 196.

CHAPTER NINE

1 Wolfgang Pauli, in a letter to Niels Bohr, February 11, 1924, as quoted by Lindley, David, 2007, p. 109.

2 Frank, Philipp, 1941, p. 7.

3 Ibid., p. 11.

4 Max Born and Pascual Jordan, 1925, as quoted by Thorndike Greenspan, Nancy, 2005, p. 124.

5 Cassidy, David C., 1991, p. 184.

6 Ibid., pp. 184–85.

7 Ibid., pp. 14–15.

8 Ibid., p. 12.

9 Ibid., p. 11.

10 Ibid., p. 15.

11 Ibid., p. 140.

12 Max Born, 1923, as quoted by Lightman, Alan, 2005.

13 Thorndike Greenspan, Nancy, 2005, p. 116.

14 Born, Gustav, 2002.

15 Max Born, in a letter to Albert Einstein, April 7, 1923, in Einstein, Albert, and Max Born, 1971, p. 75.

16 Born, Gustav, 2002, p. 245.

17 Cassidy, David C., 1991, pp. 148–49.

18 Ibid., p. 152.

19 Ibid., pp. 171–72.

20 Heisenberg, Werner, 1971, p. 46.

21 Pais, Abraham, 1991, p. 263.

22 Cassidy, David C., 1991, p. 189.

23 Schroer, Bert, 2005.

24 Mehra, Jagdish, and Helmut Rechenberg, 2001, p. 56.

25 Schroer, Bert, 2005.

26 Ibid.

27 Schroer, Bert, 2007, p. 5.

28 Paul Dirac, as quoted by Kragh, Helge, 1990, p. 8.

29 Dalitz, R.H., and Rudolph Peierls, 1986.

30 Kragh, Helge, 1990, p. 3.

31 Ibid., p. 4.

32 Ibid., p. 10.

33 Pais, Abraham, 1998, p. 5.

34 Kragh, Helge, 1990, p. 10.

35 Dirac, Monica, 2002.

36 Dalitz, R.H., and Rudolph Peierls, 1986, p. 146.

37 The title of Prince is not a French title, unlike Duc, which takes precedence over it. The title of Prince or Princess could be used by the descendants of a Duc. Maurice de Broglie, for instance, was a Prince until his father died and he inherited the title of Duc.

38 Louis de Broglie, as quoted by Abragam, Anatole, 1988, p.28.

39 Ibid., p. 26.

40 Maurice de Broglie, as quoted by Abragam, Anatole, 1988, p. 27.

41 Raman, V.V., and Paul Forman, 1969, p. 296.

42 Klein, Martin J., 1964c, p. 32.

43 Louis de Broglie, in a letter to Abraham Pais, August 9, 1978, as quoted by Pais, Abraham, 1982, p. 438. Brackets indicate a section translated from French.

CHAPTER TEN

1 Frank, Philipp, 1947, *Einstein, His Life and Times*, tr. George Rosen, Alfred A, Knopf, p. 117.

2 Albert Einstein, in a letter to Paul Ehrenfest, July 12, 1924, as quoted by Pais, Abraham, 1982, p. 425.

3 Albert Einstein, in a footnote to S.N. Bose, 1924, "Planck's Law and the Hypothesis of Light Quanta," as quoted by Klein, Martin J.,1964c, p. 26.

4 Abragam, Anatole, 1988, p. 30.

5 Albert Einstein, 1925, as quoted by Klein, Martin J., 1964c, p. 33.

6 Paul Langevin, as quoted by Abragam, Anatole, 1988, p. 31.

7 Ibid.

8 Raman, V.V., and Paul Forman, 1969, p. 301.

9 Klein, Martin J., 1964c, p. 32.

10 Elsasser, Walter, 1978, pp. 59–61.

11 Ibid., p. 61.

12 Max Born, in a letter to Albert Einstein, July 15, 1925, in Einstein, Albert, and Max Born, 1971, p. 83.

13 There is some debate about Walter Elsasser's version of events regarding his 1925 note on de Broglie matter waves and that of Max Born, who has said he and James Franck hit upon the connection between the Bell Lab experiment and matter waves and then referred the matter to Elsasser. According to Elsasser, he recalls very well what happened because it was his very first academic publication and a big event in his life. Born was under considerable stress that summer, and may simply have confused Elsasser with Friedrich Hund, the student he had referred the problem to earlier in the year. Elsasser also says he did not inform Born of what he was doing, only Franck. But Born's letter to Einstein on July 15, 1925, indicates he was aware of Elsasser's work prior to the paper being sent off for publication on July 18; Born was most likely kept informed by Franck.

14 Mehra, Jagdish, and Helmut Rechenberg, 2001, p. 374. Walter Elsasser completed his paper on July 18, 1925.

15 Walter Elsasser, 1925, as quoted by Klein, Martin J., 1964c, p. 39.

16 Rubin, Harry, 1995, p. 112.

17 For an attempt to "fill in the gaps" in the calculations in Heisenberg's seminal 1925 paper, see Aitchison, Ian J.R., David A. MacManus and Thomas M. Snyder, 2004.

18 Bernstein, Jeremy, 2005.

19 Ibid., p. 1003.

20 Max Born, in a letter to Albert Einstein, July 15, 1925, in Einstein, Albert, and Max Born, 1971, p. 84.

21 Born, Max, 1969, p. 92.

22 Bowen, Marshall, and Joseph Coster, 1980.

23 Wolfgang Pauli, in a letter to Ralph Kronig, October 9, 1925, as quoted by Cassidy, David C., 1992, p. 204.

24 Wolfgang Pauli, as quoted by Thorndike Greenspan, Nancy, 2005, pp. 125–26.

25 Mehra, Jagdish, and Helmut Rechenberg, 2001, p. 62.

26 Thorndike Greenspan, Nancy, 2005, p. 126.

27 Pascual Jordan, as quoted by Mehra, Jagdish, and Helmut Rechenberg, 2001, p. 63.

28 Mehra, Jagdish, and Helmut Rechenberg, 2001, p. 64.

29 Niels Bohr, as quoted by Pais, Abraham, 1991, pp. 279–80.

30 Niels Bohr, in a letter to Paul Ehrenfest, October 14, 1925, as quoted by Pais, Abraham, 1991, p. 280.

31 Werner Heisenberg, in a letter to Wolfgang Pauli, October 12, 1925, as quoted by Lindley, David, 2007, p. 125.

32 Max Born, in a letter to Niels Bohr, October 10, 1925, as quoted by Thorndike Greenspan, Nancy, 2005, p. 128.

33 Niels Bohr, in a letter to Carl Oseen, January 29, 1926, as quoted by Thorndike Greenspan, Nancy, 2005, p. 128.

34 Thorndike Greenspan, Nancy, 2005, p. 135.

CHAPTER ELEVEN

1 Goudsmit, Samuel, 1971.

2 Paul Ehrenfest, in a quote as recalled by Goudsmit, Samuel, 1971.

3 Goudsmit, Samuel, 1971.

4 Klein, Martin J., 1986, p. 337.

5 Pais, Abraham, 1991, p. 243.

6 Niels Bohr, in a letter to Ralph Kronig, March 26, 1926, as quoted by Klein, Martin J., 1986, p. 338.

7 Paul Ehrenfest, in a letter to Albert Einstein, September 16, 1925, as quoted by Klein, Martin J., 1986, p. 336.

8 Goudsmit, Samuel, 1971.

9 Albert Einstein, in a letter to Arnold Sommerfeld in early 1922, as quoted by Eckert, M., 2000, p. 152.

10 Eckert, M., 2000, p. 152.

11 Cline, Barbara Lovett, 1965, pp. 129–31.

12 Beller, Mara, 1999b, p. 259.

13 Honner, John, 1982.

14 Beller, Mara, 1999b, p. 257.

15 Frank, Philipp, 1947, p. 117.

16 Ibid., p. 119.

17 Ibid., p. 207.

18 Albert Einstein, as quoted by Frank, Philipp, 1947, p. 206.

19 *Physique du monde*, 1780, as quoted by Gingras, Yves, 2001, p. 402.

20 Niels Bohr, in a letter to Paul Ehrenfest, December 22, 1925, as quoted by Klein, Martin J., 1986, p. 338.

21 Max Born, in a letter to Niels Bohr, December 1925, as quoted by Thorndike Greenspan, Nancy, 2005, p. 136.

22 Kragh, Helge, 1990, p. 14.

23 Werner Heisenberg, in a letter to Paul Dirac, November 20, 1925, as quoted by Kragh, Helge, 1990, p. 20.

24 Kragh, Helge, 1990, p. 20.

25 Max Born, as quoted by Beller, Mara, 1999a, p. 31.

CHAPTER TWELVE

1 Moore, Walter, 1989, p. 168.

2 Ibid., pp. 129, 188.

3 Raman, V.V., and Paul Forman, 1969, p. 297.

4 Moore, Walter, 1989, p. 12.

5 Ibid., p. 31.

6 Erwin Schrödinger, in a letter to Arthur Eddington, March 22, 1940, as quoted by Moore, Walter, 1989, p. 41.

7   Moore, Walter, 1989, pp. 64–65.

8   Ibid., pp. 65–66.

9   Erwin Schrödinger, in a letter to Wolfgang Pauli, November 8, 1922, as quoted by Moore, Walter, 1989, p. 152.

10 Erwin Schrödinger, in his inaugural lecture, December 9, 1922, as quoted by Moore, Walter, 1989, p. 153.

11 Moore, Walter, 1989, p. 163.

12 Erwin Schrödinger, in a letter to Albert Einstein, November 3, 1925, as quoted by Moore, Walter, 1989, p. 192.

13 Bloch, Felix, 1976.

14 Schrödinger, Erwin, 1925, "On Einstein's Theory of Gas," as quoted by Klein, Martin J., 1964c, p. 43.

15 Niels Bohr, in a letter to Ernest Rutherford, January 27, 1926, as quoted by Kalckar, Jørgen, 1985.

16 Beller, Mara, 1999a, p. 140.

17 Albert Einstein, in a letter to Michele Besso, December 25, 1925, as quoted by Kalckar, Jørgen, 1985, p. 8.

18 Thorndike Greenspan, Nancy, 2005, p. 131.

19 Einstein, Alfred, and Max Born, 1971, p. 89.

20 Llewellyn Thomas, in a letter to Samuel Goudsmit, March 25, 1926, as quoted by Goudsmit, Samuel, 1971.

21 Mehra, Jagdish, and Helmut Rechenberg, 2001, p. 618.

22 Erwin Schrödinger, in a letter to Willy Wien, February 22, 1926, as quoted by Mehra, Jagdish, and Helmut Rechenberg, 2001, p. 639.

23 Wolfgang Pauli, in a letter to Pascual Jordan, April 12, 1926, as quoted by Mehra, Jagdish, and Helmut Rechenberg, 2001, p. 656.

24 Niels Bohr, in a letter to Werner Heisenberg, April 10, 1926, as quoted by Mehra, Jagdish, and Helmut Rechenberg, 2001, p. 655.

25 Thorndike Greenspan, Nancy, 2005, p. 138.

26 Paul Dirac, as quoted by Kragh, Helge, 1990, p. 31.

27 Albert Einstein, in a letter to Erwin Schrödinger, April 14, 1926, as quoted by Moore, Walter, 1989, p. 209.

28 Paul Ehrenfest, in a letter to Erwin Schrödinger, May 19, 1926, as quoted by Moore, Walter, 1989, p. 210.

CHAPTER THIRTEEN

1   Beller, Mara, 1996. See also de Regt, Heck W., 1991.

2   Erwin Schrödinger, as quoted by Beller, Mara, 1999a, p. 32.

3   Moore, Walter, 1989, p. 207.

4   de Regt, Heck W., 1991, p. 476.

5   Erwin Schrödinger, as quoted by Moore, Walter, 1989, p. 208.

6   Cassidy, David C., 1992, pp. 216–17.

7   Beller, Mara, 1999a, pp. 30–31.

8   Bloch, Felix, 1976, p. 24.

9   Hartmut Kallmann, as quoted by Mehra, Jagdish, and Helmut Rechenberg, 2001, p. 636. The term "[another physicist]" refers to Walter Gordon, a doctoral student at the University of Berlin.

10 Werner Heisenberg, in a letter to Paul Dirac, May 26, 1926, as quoted by Kragh, Helge, 1991, p. 32.

11 Kragh, Helge, 1991, p. 32.

12 Beller, Mara, 1999a, p. 30.

13 See Beller, Mara, 1996, and de Regt, Heck W., 1991.

14 Erwin Schrödinger, as quoted by Moore, Walter, 1989, p. 219.

15 Moore, Walter, 1989, p. 222.

16 Ibid.

17 Cassidy, David C., 1992, p. 152.

18 Moore, Walter, 1989, p. 222.

19 Beller, Mara, 1990, p. 567.

20 Ibid., p. 574.

21 Niels Bohr, as quoted by Moore, Ruth, 1966, p. 143.

22 Niels Bohr, as quoted by Moore, Ruth, 1966, p. 144.

23 Pascual Jordan, in a letter to Niels Bohr, July 29, 1926, as quoted by Mehra, Jagdish, and Helmut Rechenberg, 2001, p. 56.

24 Ibid.

25 Halpern, Paul, 2004.

26 George Uhlenbeck, as quoted by Pais, Abraham, 1982, p. 332.

27 Dalitz, R.H., and Rudolph Peierls, 1986, p. 146.

28 Pascual Jordan, as quoted by Kragh, Helge, 1991, p. 25.

29 Paul Ehrenfest, in a letter to Paul Dirac, October 1, 1926, as quoted by Kragh, Helge, 1991, p. 53.

30 Max Delbruck, as quoted by Thorndike Greenspan, Nancy, 2005, p. 141.

31 Elsasser, Walter, 1978, pp. 68–71.

CHAPTER FOURTEEN

1 Moore, Ruth, 1966, p. 146.

2 Werner Heisenberg, as quoted by Pais, Abraham, 1991, pp. 298–99.

3 Moore, Walter, 1989, p. 250.

4 Niels Bohr, in a letter to Ralph Fowler, October 26, 1926, as quoted by Pais, Abraham, 1991, p. 300.

5 Erwin Schrödinger, in a letter to Niels Bohr, October 23, 1926, as quoted by de Regt, Heck W., 1991.

6 Beller, Mara, 1999a, p. 47.

7 Ibid., p. 68.

8 Paul Dirac, as quoted by Kragh, Helge, 1990, p. 37.

9 Victor Weisskopf, as quoted by Beller, Mara, 1999b.

10 Beller, Mara, 1999b, p. 261.

11 Kragh, Helge, 1999, p. 37.

12 Moore, Walter, 1989, pp. 224–25.

13 Ibid., p. 231.

14 Peierls, R.E., 1960, p. 184.

15 Beller, Mara, 1990, pp. 575–76.

16 Wolfgang Pauli, in a letter to Werner Heisenberg, October 19, 1926, as quoted by Moore, Walter, 1989, p. 232.

17 Beller, Mara, 1999a, p. 126.

18 Lindley, David, 2007, p. 142.

19 Oskar Klein, as quoted by Pais, Abraham, 1991, p. 303.

20 Beller, Mara, 1999a, p. 89.

21 Wolfgang Pauli, in a letter to Werner Heisenberg, October 19, 1926, as quoted by Lindley, David, 2007, p. 145.
22 Beller, Mara, 1999a, pp. 92, 95.
23 Ibid., footnote, p. 20.
24 Ibid., p. 105.
25 Ibid., p. 107.
26 Werner Heisenberg, in a letter to Wolfgang Pauli, March 9, 1927, as quoted by Beller, Mara, 1999a, p. 110.

CHAPTER FIFTEEN

1 Bacciagaluppi, Guido, and Antony Valentini, (publication pending), p. 10.
2 Moore, Walter, 1989, p. 232. See also Bacciagaluppi, Guido, and Antony Valentini, (publication pending), footnote, p. 11.
3 Gehrenbeck, Richard K., 1978.
4 Bacciagaluppi, Guido, and Antony Valentini, (publication pending), p. 44.
5 Ibid., p. 72.
6 Beller, Mara, 1996, p. 552.
7 Werner Heisenberg, as quoted by Pais, Abraham, 1991, p. 310.
8 Cassidy, David C., 1992, p. 243.
9 Hendry, John, 1984, p. 124.
10 Beller, Mara, 1991a, p. 140.
11 Ibid., p. 33.
12 Bacciagaluppi, Guido, and Antony Valentini, (publication pending), p. 19.
13 Albert Einstein, in a letter to Hendrik Lorentz, June 17, 1927, as quoted by Pais, Abraham, 1982, p. 432.
14 Werner Heisenberg, in a letter to Albert Einstein, June 10, 1927, as quoted by Pais, Abraham, 1982, p. 467.
15 Niels Bohr, in a letter to Albert Einstein, April 13, 1927, as quoted by Honner, John, 1982, p. 8.
16 Albert Einstein, as quoted by Pais, Abraham, 1982, p. 443.
17 Albert Einstein, as quoted by Howard, Don, 2005, p. 9.
18 Howard, Don, 2005, p. 12.
19 Bacciagaluppi, Guido, and Antony Valentini, (publication pending), p. 93.
20 Pais, Abraham, 1991, p. 311.
21 Ibid.
22 Wolfgang Pauli, in a letter to Niels Bohr, August 6, 1927, as quoted by Bacciagaluppi, Guido, and Antony Valentini, (publication pending), p. 61.
23 Pais, Abraham, 1991, p. 316.
24 Beller, Mara, 1999a, p. 8.
25 Ibid., p. 141.
26 Beller, Mara, 1996, p. 556.
27 Cassidy, David C., 1992, p. 246.
28 Bacciagaluppi, Guido, and Antony Valentini, (publication pending), p. 12.
29 Frans van Lunteren, personal communication, 2007.
30 Elsasser, Walter, 1978, pp. 73–74.
31 Ibid., pp. 82–83.
32 Ibid., pp. 84, 90.
33 Ibid., p. 91.
34 Ibid., pp. 91–92.

CHAPTER SIXTEEN

1 Bacciagaluppi, Guido, and Antony Valentini, (publication pending), p. 19.
2 Ibid., p. 21.
3 Werner Heisenberg, as quoted by Pais, Abraham, 1991, p. 303.
4 Howard, Don, 2005.
5 Paul Ehrenfest, in a letter to Samuel Goudsmit and George Uhlenbeck, November 3, 1927, as quoted by Segrè, Gino, 2007, p. 158.
6 For details, see the following discussions in Bacciagaluppi, Guido, and Antony Valentini, (publication pending): de Broglie discussions, pp. 399–406; Heisenberg and Born discussions, pp. 442–44; Schrödinger discussions, pp. 469–74.
7 Max Born and Werner Heisenberg, as quoted by Thorndike Greenspan, Nancy, 2005, pp. 147–48.
8 Max Born, as quoted by Cassidy, David C., 1992, p. 250.
9 Nancy Thorndike Greenspan discovered the film footage taken at the 1927 Solvay Conference by Irving Langmuir among Max Born's papers when she was researching her biography of Born. It can be viewed online at www.maxborn.net.
10 Moore, Ruth, 1966, p. 164.
11 Bacciagaluppi, Guido, and Antony Valentini, (publication pending), p. 488.
12 Ibid., p. 489.
13 Ibid., footnote, p. 21.
14 Paul Ehrenfest, as quoted by Bacciagaluppi, Guido, and Antony Valentini, (publication pending), p. 21.
15 Bacciagaluppi, Guido, and Antony Valentini, (publication pending), p. 25.
16 Hendrik Lorentz, as quoted by Bacciagaluppi, Guido, and Antony Valentini, (publication pending), p. 478.
17 Werner Heisenberg, as quoted by Pais, Abraham, 1991, p. 320.
18 Werner Heisenberg, as quoted by Blair Bolles, Edmund, 2004, p. 274.
19 Albert Einstein, as quoted by Blair Bolles, Edmund, 2004, p. 275.
20 Moore, Walter, 1989, pp. 241, 249.
21 Bloch, Felix, 1976, p. 25.
22 Jacobi, Manfred, 2000.
23 Segrè, Gino, 2007, p. 159.
24 Kragh, Helge, 1990, p. 57.
25 Ibid., p. 49.
26 Paul Dirac, as quoted by Kragh, Helge, 1990, p. 61.
27 Pais, Abraham, 1991, p. 316.
28 Bacciagaluppi, Guido, and Antony Valentini, (publication pending), p. 23.
29 Blair Bolles, Edmund, 2004, pp. 226–27.
30 Richardson, O.W., 1929.
31 Holton, Gerald, 2004.
32 Ibid., p. 13.
33 Neffe, Jürgen, 2007, p. 196.
34 Thorndike Greenspan, Nancy, 2005, pp. 151–53.
35 Max Born, in a letter to Norbert Wiener, January 10, 1929, as quoted by Thorndike Greenspan, Nancy, 2005, p. 153.
36 Pauline de Pange, as quoted by Abragam, Anatole, 1988, p. 39.
37 Abragam, Anatole, 1988, p. 39.
38 Pais, Abraham, 1991, p. 311.
39 Bacciagaluppi, Guido, and Antony Valentini, (publication pending), p. 23.

40 Werner Heisenberg, as quoted by Bacciagaluppi, Guido, and Antony Valentini, (publication pending), p. 24.

CHAPTER SEVENTEEN

1   Frank, Philipp, 1947, p. 219.
2   Pais, Abraham, 1982, p. 346.
3   Frank, Philipp, 1947, p. 219.
4   Pais, Abraham, 1982, p. 346.
5   Ibid., p. 247.
6   According to science historian Martin J. Klein, who reviewed all the Ehrenfest correspondence while it was still in the Ehrenfest home, there were some letters that indicated Einstein would be going to talk to Tatyana in the summer of 1932, but there are no details of what transpired on that visit.
7   Hugo Ehrenfest, in a letter to Paul Ehrenfest, October 2, 1931, as quoted by Halpern, Paul, 2006.
8   Martin J. Klein, personal communication, March 2007.
9   Paul Ehrenfest, in a letter to Niels Bohr, May 13, 1931, as quoted by Pais, Abraham, 1991, p. 409.
10 Martin J. Klein, personal communication, March 2007.
11 Ibid.
12 Paul Ehrenfest, in a letter dated August 14, 1932, but not sent, as quoted by Pais, Abraham, 1991, p. 409.
13 Letter to Paul Ehrenfest, September 30, 1932, Ehrenfest Personal Correspondence, #4, Section 10, Paul Ehrenfest Archive, Museum Boerhaave, Leiden.
14 Martin J. Klein, personal communication, March 2007.
15 Thorndike Greenspan, Nancy, 2005, p. 155.
16 Ibid., pp. 154–55.
17 Ibid., p. 158.
18 Beller, Mara, 1999a, p. 38.
19 Enz, Charles P., 2000, p. 73.
20 Funk, Herbert, 2000, p. 28.
21 Segrè, Gino, 2007, p. 199.
22 Ibid., p. 201.
23 Enz, Charles P., 2000, pp. 73–74.
24 Moore, Walter, 1989, p. 255.
25 Cassidy, David C., 1992, p. 289.
26 Ibid., p. 265.
27 Niels Bohr, in a letter to Wolfgang Pauli, July 1, 1929, as quoted by Honner, John, 1982, p. 4.
28 Wolfgang Pauli, in a letter to Niels Bohr, July 17, 1929, as quoted by Honner, John, 1982, p. 3.
29 Cassidy, David C., 1992, p. 257.
30 Beller, Mara, 1999a, p. 11.
31 Beller, Mara, 1999b, pp. 255, 257.
32 Pais, Abraham, 1991, pp. 333–34, 226.
33 Taylor, Simon, 1983, pp. 123–24.
34 Levenson, Thomas, 2003, p. 419.
35 Taylor, Simon, 1983, p. 124.

36 Max Born, as quoted by Thorndike Greenspan, Nancy, 2005, p. 177.

37 Ibid., p. 176.

38 Beyler, Richard H., 1996, p. 259.

39 Moore, Walter, 1989, pp. 273–74.

40 Thorndike Greenspan, Nancy, 2005, p. 180.

41 Moore, Walter, 1989, p. 273.

42 Erwin Schrödinger, in a letter to Albert Einstein, August 12, 1933, as quoted by Moore, Walter, 1989, pp. 273–75.

43 Frank, Philipp, 1941, pp. 240–41.

44 Cornwell, John, 2003, p. 34.

45 Thorndike Greenspan, Nancy, 2005, p. 181.

46 Hugo Ehrenfest, in a letter to Paul Ehrenfest, June 1933, as quoted by Halpern, Paul, 2006.

47 Martin J. Klein, personal communication, March 2007.

48 Thorndike Greenspan, Nancy, 2005, p. 182.

49 Ogilvie, Marilyn Bailey, and Joy Dorothy Harvey, eds., 2000, p. 407.

50 Kragh, Helge, 1991, p. 107.

51 Segrè, Gino, 2007, p. 251.

52 Paul Ehrenfest, as recalled by Paul Dirac in a letter to Niels Bohr, September 28, 1933, as quoted by Pais, Abraham, 1991, p. 410.

53 Martin J. Klein, personal communication, March 2007.

54 Ibid. There are a number of conflicting stories about the circumstances surrounding Vassily's death, such as whether he was merely blinded when his father shot him; whether he was killed outright inside the Professor Watering Institute; or whether the murder-suicide took place in the large, busy park outside the institution. Klein notes that he and Abraham Pais discussed the question of Vassily's death at length and concluded that it seems highly unlikely that such an incident (for which there were no witnesses) would have gone unnoticed in the park, but that it is possible Vassily was not killed outright at the time of the shooting.

55 Einstein, Albert, 1950, "Paul Ehrenfest in Memoriam," *Out of My Later Years*, Philosophical Library, pp. 238, 239.

56 Neffe, Jürgen, 2007, pp. 33, 267.

57 Klein, Martin J., 1981, p. 3.

CHAPTER EIGHTEEN

1 Peter Galison, "History," in Gross, David, Marc Hanneaux and Alexander Sevrin, eds., 2007, p. 17.

2 David Gross, as quoted in "Nobel Laureate admits string theory in trouble," *New Scientist*, 188 (2959), December 10, 2005.

3 Albert Einstein, as quoted by Mehra, Jagdish, 1975, p. xiv.

4 Paul Langevin, as quoted by Bacciagaluppi, Guido, and Antony Valentini, (publication pending), p. 25.

5 David Gross, as quoted by Harris, Richard, August 27, 2007.

EPILOGUE

1 Ogilvie, Marilyn Bailey, and Joy Dorothy Harvey, eds., 2000, p. 407.

BIBLIOGRAPHY

Abragam, Anatole, 1988, "Louis Victor Pierre Raymond de Broglie, 15 August 1892–19 March 1987," *Biographical Memoirs of Fellows of the Royal Society*, 34, pp. 22–41.

Aitchison, Ian J.R., David A. MacManus and Thomas M. Snyder, 2004, "Understanding Heisenberg's 'Magical' Paper of July 1925: A New Look at the Calculational Details," *American Journal of Physics*, 72, pp. 1370–79.

Bacciagaluppi, Guido, and Antony Valentini, (publication pending), *Quantum Theory at the Crossroads: Reconsidering the 1927 Solvay Conference*, Cambridge University Press. The full text is available on-line at http://arvix.org/abs/quant-ph/0609184v1.

Badash, Lawrence, 1972, "The Completeness of Nineteenth-Century Science," *Isis*, 63, No. 1.

Beller, Mara, 1990, "Born's Probabilistic Interpretation: A Case Study of 'Concepts in Flux,'" *Studies in History and Philosophy of Science*, 21, No. 4, pp. 563–88.

Beller, Mara, 1996, "The Conceptual and the Anecdotal History of Quantum Mechanics," *Foundations of Physics*, 26, No. 4, pp. 545–57.

Beller, Mara, 1999a, *Quantum Dialogue: The Making of a Revolution*, University of Chicago Press.

Beller, Mara, 1999b, "Jocular Commemorations: The Copenhagen Spirit," *Osiris*, 14, pp. 252–73.

Bernstein, Jeremy, 2005, "Max Born and the Quantum Theory," *American Journal of Physics*, 73, pp. 999–1008.

Beyler, Richard H., 1996, "Targeting the Organism: The Scientific and Cultural Context of Pascual Jordan's Quantum Biology, 1932–1947," *Isis*, 87, pp. 248–73.

Blackmore, John, 1985, "An Historical Note on Ernst Mach," *British Journal for the Philosophy of Science*, 36, pp. 299–305.

Blair Bolles, Edmund, 2004, *Einstein Defiant: Genius versus Genius in the Quantum Revolution*, Joseph Henry Press.

Bloch, Felix, 1976, "Heisenberg and the Early Days of Quantum Mechanics," *Physics Today*, 29, pp. 23–27.

Born, Gustav, 2002, "The Wide-Ranging Family History of Max Born," *Notes and Recordings of the Royal Society of London*, 56, No. 2, pp. 232–33.

Born, Max, 1969, *Physics in My Generation*, Springer Verlag.

Boutin, Paul, 2005, "Theory of Anything? Physicist Lawrence Krauss Turns on His Own," *Slate*, November 23, 2005. Available online at www.slate.com/id/2131014/.

Bowen, Marshall, and Joseph Coster, 1980, "Born's Discovery of the Quantum-Mechanical Matrix Calculus," *American Journal of Physics*, 48, pp. 491–92.

Boya, Luis J., 2003, "Rejection of the Light Quantum: The Dark Side of Niels Bohr," *International Journal of Theoretical Physics*, 42, No. 10, pp. 2563–73.

Bibliography

Brush, Stephen G., 1980, "The Chimerical Cat: Philosophy of Quantum Mechanics in Historical Perspective," *Social Studies of Science*, 10, No. 4, pp. 339–447.

Cassidy, David C., 1992, *Uncertainty: The Life and Science of Werner Heisenberg*, W.H. Freeman and Company.

Cline, Barbara Lovett, 1965, *Men Who Made a New Physics*, University of Chicago Press.

Comte, Auguste, 1842, *A General View of Positivism*, 1957, tr. J.H. Bridges, Robert Speller and Sons.

Cornwell, John, 2003, *Hitler's Scientists*, Viking.

Curtis, Theodor, 1961, "Robert Bunsen," in *Great Chemists*, ed. Eduard Farber, Interscience.

Dalitz, R.H., and Rudolph Peierls, 1986, "Paul Adrien Maurice Dirac, 8 August 1902–20 October 1984," *Biographical Memoirs of Fellows of the Royal Society*, 32, pp. 138–85.

de Regt, Heck W., 1991, "Erwin Schrödinger, *Anschaulichkeit*, and Quantum Theory," *Studies in History and Philosophy of Modern Science*, 28, No. 4, pp. 461–81.

Dirac, Monica, 2002, on the occasion of the centennial celebration of Paul Dirac's career at Cambridge University, http://www.damtp.cam.ac.uk/strings02/dirac/dirac/.

Dresden, Max, 1987, *H.A. Kramers—Between Tradition and Revolution*, Springer Verlag.

Ebeling, Werner, and Dieter Hoffman, 1991, "The Berlin School of Thermodynamics Founded by Helmholtz and Clausius," *European Journal of Physics*, 12, pp. 1–9.

Eckert, M., 2000, "The Emergence of Quantum Schools: Munich, Göttingen and Copenhagen as New Centers of Atomic Theory," *Annalen der Physik*, 10, No. 1–2, pp. 151–62.

Einstein, Albert, 1919, tr. Bertram Schwarzschild, 2005, "Albert Einstein to Paul Ehrenfest," *Physics Today*, 58, No. 4, p. 88.

Einstein, Albert, 1959, "Autobiographical Notes," in *Albert Einstein: Philosopher-Scientist*, ed. Paul Arthur Schlipp, Harper & Brothers.

Einstein, Albert, 1995, *The Collected Papers of Albert Einstein*, Vol. 5, *The Swiss Years: Correspondence, 1902–1914*, tr. Anna Beck, Princeton University Press.

Einstein, Albert, and Max Born, 1971, *The Born–Einstein Letters*, tr. Irene Born, Walker and Company.

Elsasser, Walter, 1978, *Memories of a Physicist in the Atomic Age*, Science History Publications.

Enz, Charles P., 2000, "Wolfgang Pauli–Carl Gustav Jung, a Dialogue over the Boundaries," *Wolfgang Pauli and Modern Physics*, ETH-Bibliothek.

Frank, Philipp, 1941, *Between Physics and Philosophy*, Harvard University Press.

Funk, Herbert, 2000, "Wolfgang Pauli—A Biographical Sketch," *Wolfgang Pauli and Modern Physics*, ETH-Bibliothek.

Gehrenbeck, Richard K., 1978, "Electron Diffraction: Fifty Years Ago," *Physics Today*, 31, pp. 34–41.

Gershenson, Daniel E., and Daniel A. Greenberg, 1964, "The Physics of the Eleatics," in *The Natural Philosopher*, Vol. 3, ed. Daniel E. Gershenson and Daniel A. Greenberg, Blaisdell Publishing Company, pp. 99–111.

Gingras, Yves, 2001, "What Did Mathematics Do to Physics?" *History of Science*, 39, pp. 383–416.

Goudsmit, Samuel, 1971, "The Discovery of the Electron Spin, a Lecture Presented to the Dutch Physical Society." The transcript is available online at http://www.lorentz.leidenuniv.nl/history/spin/goudsmit.html.

Greene, Brian, 2000, *The Elegant Universe*, Vintage Books.

Gross, David, Marc Hanneaux and Alexander Sevrin, eds., 2007, *The Quantum Structure of Space and Time: Proceedings of the 23rd Solvay Conference on Physics, Brussels, Belgium, 1–3 December 2005*, World Scientific Publishing Company.

Halpern, Paul, 2004, "Nordström, Ehrenfest and the Role of Dimensionality in Physics," *Physics in Perspective*, 6, No. 4, pp. 390–400.

Halpern, Paul, 2006, "Brotherly Advice: Letters from Hugo to Paul Ehrenfest in His Final

Years," presented at the American Physical Society March Meeting, March 16, 2006, Baltimore, Maryland.

Harris, Richard, 2007, "Short of 'All,' String Theorists Accused of Nothing," National Public Radio, August 27, 2007. Available online at www.npr.org/templates/story/story.php?storyID=6377252.

Heilbron, John L., and Thomas S. Kuhn, 1969, "The Genesis of the Bohr Atom," in *Historical Studies in the Physical Sciences*, ed. Russell McCormmach, University of Pennsylvania Press.

Heisenberg, Werner, 1971, *Physics and Beyond: Encounters and Conversations*, Harper and Row.

Hendry, John, 1984, *The Creation of Quantum Mechanics and the Bohr–Pauli Dialogue*, D. Reidel Publishing Company.

Highfield, Roger, and Paul Carter, 1993, *The Private Lives of Albert Einstein*, Faber and Faber.

Hoffman, Banesh, with Helen Dukus, 1972, *Albert Einstein: Creator and Rebel*, Plume (Penguin Books).

Holton, Gerald, 1992, "Ernst Mach and the Fortunes of Positivism in America," *Isis*, 83, No. 1, pp. 27–60.

Holton, Gerald, 2003, "Einstein's Third Paradise," *Daedalus*, 132, pp. 26–34.

Holton, Gerald, 2004, "Paul Tillich, Albert Einstein, and the Quest for the Ultimate," The Paul Tillich Lecture, Harvard University. This lecture is available online at http://www.physics.harvard.edu/holton/Tillich.pdf.

Honner, John, 1982, "The Transcendental Philosophy of Niels Bohr," *Studies in History and Philosophy of Science*, 13, No. 1, pp. 1–29.

Howard, Don, 2005, "Revisiting the Einstein–Bohr Dialogue," from a Bar-Hillel Lecture given in Jerusalem, University of Notre Dame. This lecture is available online at www.nd.edu/~dhoward1/papers.html.

Huijnen, Pim, and A.J. Knox, 2007, "Paul Ehrenfest's Rough Road to Leiden: A Physicist's Search for a Position, 1904–1912," *Physics in Perspective*, 9, pp. 186–211.

Jacobi, Manfred, 2000, "Wolfgang Pauli's Family Background," *Gesnerus*, 57, Nos. 3–4.

James, William, 1926, *The Letters of William James*, ed. Henry James, Little Brown.

Johnson, George, 2007, "Meta Physicists," *The New York Times Book Review*, June 24, 2007.

Jordan, Pascual, 1975, "My Recollections of Wolfgang Pauli," tr. Ira M. Freeman, *American Journal of Physics*, 43, No. 3, pp. 205–8.

Kalckar, Jørgen, 1985, *Foundations of Quantum Mechanics I (1926–1932)*, North-Holland Publishing Company, pp. 456–57.

Klein, Étienne, 1997, *Conversations with a Sphinx: Paradoxes in Physics*, tr. David Le Vay, Souvenir Press.

Klein, Martin J., 1964a, "Einstein's First Paper on Quanta," in *The Natural Philosopher*, Vol. 11, ed. Daniel E. Gershenson and Daniel A. Greenberg, Blaisdell Publishing Company, pp. 59–86.

Klein, Martin J., 1964b, "Planck, Entropy and Quanta, 1901–1906," in *The Natural Philosopher*, Vol. 1, ed. Daniel E. Gershenson and Daniel A. Greenberg, Blaisdell Publishing Company, pp. 83–108.

Klein, Martin J., 1964c, "Einstein and Wave-Particle Duality," in *The Natural Philosopher*, Vol. 3, ed. Daniel E. Gershenson and Daniel A. Greenberg, Blaisdell Publishing Company, pp. 3–49.

Klein, Martin J., 1970a, *Paul Ehrenfest*, Vol. 1, *The Making of a Theoretical Physicist*, North-Holland Publishing Company.

Klein, Martin J., 1970b, "The First Phase of the Bohr–Einstein Dialogue," *Historical Studies in the Physical Sciences*, Vol. 2, pp. 1–39.

Klein, Martin J., 1981, "Not By Discoveries Alone: The Centennial of Paul Ehrenfest," *Physica*, 106A, pp. 3–14.

Bibliography

Klein, Martin J., 1986, "Great Connections Come Alive: Bohr, Ehrenfest and Einstein," in *The Lessons of Quantum Theory*, eds. J. de Boer, E. Dal, and O. Ulfbeck, Elsevier Science Publishers.

Klein, Martin J., and Allan Needell, 1977, "Some Unnoticed Publications by Einstein," *Isis*, 68, No. 4, pp. 601–4.

Kragh, Helge, 1990, *Dirac: A Scientific Biography*, Cambridge University Press.

Kragh, Helge, 1999, *Quantum Generations*, Princeton University Press.

Langevin, Paul, and Maurice de Broglie, eds., 1912, *Proceedings First Solvay Conference*, Gauthier-Villars.

Levenson, Thomas, 2003, *Einstein in Berlin*, Bantam Books.

Lightman, Alan, 2005, "Metaphor in Science," chap. 3 in *A Sense of the Mysterious: Science and the Human Spirit*, Pantheon.

Lindley, David, 2007, *Uncertainty: Einstein, Heisenberg, Bohr, and the Struggle for the Soul of Science*, Doubleday.

Mach, Ernst, 1886, *The Analysis of Sensations*, English tr., 1914, 4th ed., Open Court.

Mach, Ernst, 1905, *Knowledge and Error*, tr. Erwin N. Heibert, 1976, Reidel Publishing Company.

Mehra, Jagdish, 1975, in *The Solvay Conferences of Physics*, Reidel Publishing Company.

Mehra, Jagdish, and Helmut Rechenberg, 2001, *The Historical Development of Quantum Theory*, Vol. 6, *The Completion of Quantum Mechanics, 1926–1941, Part 1: The Probabilistic Interpretation and the Empirical and Mathematical Foundation of Quantum Mechanics, 1926–1936*, Springer Verlag.

Meyenn, Karl von, and Engelbert Schucking, 2001, "Wolfgang Pauli," *Physics Today*, 54, No. 2, p. 43.

Moore, Ruth, 1966, *Niels Bohr: The Man, His Science, and the World They Changed*, Alfred A. Knopf.

Moore, Walter, 1989, *Schrödinger, Life and Thought*, Cambridge University Press.

Neffe, Jürgen, 2007, *Einstein*, tr. Shelley Frisch, Farrar, Straus and Giroux.

Ogilvie, Marilyn Bailey, and Joy Dorothy Harvey, eds., 2000, "Ehrenfest-Afanassjewa, Tatyana Alexeyevna (1876–1964)," *The Biographical Dictionary of Women in Science*, Taylor & Francis.

Ostwald, Wilhelm, 1927, *Lebenslinien—Eine Selbstbiographie*, Vol. 2, Klassing, pp. 187–88.

Pais, Abraham, 1982, *Subtle is the Lord . . . The Science and the Life of Albert Einstein*, Oxford University Press.

Pais, Abraham, 1991, *Niels Bohr's Times: In Physics, Philosophy, and Polity*, Clarendon Press.

Pais, Abraham, Maurice Jacob, David I. Olive and Michael F. Atiyah, eds., 1998, *Paul Dirac: The Man and His Work*, Cambridge University Press.

Peat, F. David, 1987, *Synchronicity*, Bantam Books.

Peierls, R.E., 1960, "Wolfgang Ernst Pauli, 1900–1958," *Biographical Memoirs of Fellows of the Royal Society*, 5, pp. 174–92.

Philip, E.B., 1956, "The Nature of Mathematics," in *The World of Mathematics*, Vol. 1, ed. James R. Newman, Simon and Shuster.

Poincaré, Henri, 1905, "The Theories of Modern Science," *Science and Hypothesis*, Walter Scott Publishing, pp. 160–82.

Quinn, Susan, 1995, *Marie Curie: A Life*, Addison Wesley.

Raman, V.V., and Paul Forman, 1969, "Why Was It Schrödinger Who Developed de Broglie's Ideas?" in *Historical Studies in the Physical Sciences*, ed. Russell McCormmach, University of Pennsylvania Press.

Reiter, Wolfgang L., 2007, "Ludwig Boltzmann: A Life of Passion," *Physics in Perspective*, 9, pp. 357–74.

Richardson, O.W., 1929, "Hendrik Antoon Lorentz," *Journal of the London Mathematical Society*, 4, No. 1, pp. 183–92.

Rigden, John S., 2005, *Einstein 1905: The Standard of Greatness*, Harvard University Press.

Rubin, Harry, 1995, "Walter M. Elsasser, Biographical Memoirs," *National Academy of Sciences*, Vol. 68, pp. 103–65.

Schroer, Bert, 2005, "Physicists in Times of War," arXiv:physics/0603095v2. Available online from the Los Alamos physics database, xxx.lanl.gov.

Schroer, Bert, 2007, "Pascual Jordan, Biographical Notes, His Contributions to Quantum Mechanics and His Role as a Protagonist of Quantum Field Theory," *Pascual Jordan (1902–1980)*, Max Planck Institute for the History of Science, Preprint 329, pp. 47–68.

Segrè, Gino, 2007, *Faust in Copenhagen: A Struggle for the Soul of Physics*, Viking.

Seppi, Ruth, 2004, "Viennese Feuilleton during the Early 1920s: Description and Analysis of Bertha Pauli's Biographical Sketches as Contributions to a Literary Genre," thesis, Brigham Young University. Available online at http://sophie.byu.edu/sophiejournal/thesis.htm.

Smolin, Lee, 2006, *The Trouble with Physics: The Rise of String Theory, the Fall of a Science, and What Comes Next*, Houghton Mifflin.

Smutný, František, 1990, "Ernst Mach and Wolfgang Pauli's Ancestors in Prague," *European Journal of Physics*, 11, pp. 257–61.

Stuewer, Roger H., 2006, "Einstein's Revolutionary Light-Quantum Hypothesis," *Acta Physica Polonica B*, 37, No. 3, pp. 543–58.

Taylor, Simon, 1983, *Germany 1918–1933*, Duckworth.

't Hooft, Gerard, 2001, "Can There Be Physics without Experiments? Challenges and Pitfalls," *International Journal of Modern Physics A*, Vol. 16, Issue 17, pp. 2895–2908.

Thorndike Greenspan, Nancy, 2005, *The End of the Certain World: The Life and Science of Max Born*, Basic Books.

Uhlenbeck, George, 1956, "Reminiscences of Professor Paul Ehrenfest," *American Journal of Physics*, 24, No. 6, pp. 431–33.

van Lunteren, Frans, 2001, "Paul Ehrenfest and Dutch Physics in the Interwar Period," a talk given at the Annual Meeting of the History of Science Society, Vancouver.

Weinberg, Steven, 1992, *Dreams of a Final Theory*, Random House.

Weinberg, Steven, 2003, "Viewpoints on String Theory," published by NOVA online at www.pbs.org/wgbh/nova/elegant/view-weinberg.html.

Woit, Peter, 2006, *Not Even Wrong: The Failure of String Theory and the Continuing Challenge to Unify the Laws of Physics*, Jonathan Cape.

Wolpert, L., 1993, *The Unnatural Nature of Science*, Harvard University.

PHOTO CREDITS